Reports on Food Safety 2005

Food Monitoring

Joint report by the Federal Government and the states (*Länder*)

(October 2006)

W0234565

Table of contents

1 Summary

Food monitoring is a system of repeated representative measurements and evaluations of levels of undesirable substances, namely pesticides, heavy metals and other contaminants, in and on foods.

Food monitoring has been carried out as two complementary analytic programmes since 2003: first, analysis of foods from a market basket[1] developed on the basis of people's consumption behaviour, with the aim to watch the residues and contamination situation under representative conditions of sampling (market basket monitoring). Second, analyses with regard to particular topical problems in the framework of particular projects (project monitoring). The market basket and the project monitoring programmes included analysis of a total of 5,159 samples of foods of domestic and foreign origin.

The following foods were selected from the market basket:

Food of animal origin
– Raw sausage (spreadable)
– Salami (air-cured)
– Carp
– Rainbow trout

Food of vegetal origin
– Rice
– Puff pastry, bread dough, muesli bars and muesli mini bars
– Linseed
– Poppy seed
– Potatoes
– Potato fritters, croquettes, potato dumpling and puree powder
– Spinach
– Artichoke
– Broccoli
– French beans
– Carrots
– Champignon, tinned/shiitake mushroom, dried
– Pear
– Peach/nectarine
– Orange
– Tangerine

– Pineapple juice/apple juice/grapefruit juice
– Partially fermented grape must, quality sparkling wine
– Marzipan raw matter, persipan raw matter, sweets from other raw material

Depending on what undesirable substances are expected, the foods were analysed for residues of plant protection products (insecticides, fungicides, herbicides) and contaminants (namely, persistent organo-chlorine compounds, musk compounds, elements, nitrate, mycotoxins, and toxic reaction products).

Project monitoring dealt with the 10 following subjects:

– Furan in bouillon and stock products, ready-to-eat meals, sauce powders, and infant food
– Carbendazim in juice of grapes, apples, pears, oranges, and in mixed juices
– Glycoside alkaloids in potatoes
– Heavy metals in preparations of vitamins, mineral substances, plant extracts and algae
– Residues of plant protection products in tomatoes
– Persistent organo-chlorine compounds and residues of plant protection products in glasshouse cucumbers
– OTA, DON and ZEA in rye and wheat flour
– Cadmium in cuttlefish products
– Benzo(a)pyrene in smoked fish
– Herbicide residues in vegetables and fresh herbs

Interpretation of findings took account of comparison with previous years, where this was possible. Yet it must be stressed that any statement and evaluation about the contamination of foods made in this report, solely refer to the foods and substances or substance groups studied in 2005.

As a whole, the findings of the 2005 food monitoring back the recommendation that nutrition should be manifold and balanced, this being a means to minimise sometimes unavoidable dietary intake of unwanted substances.

In particular, findings from market basket and project monitoring can be summarised as follows:

Food of animal origin

• **Air-cured salami** and **soft raw sausage**, such as *Teewurst* and *Mettwurst* (soft pork sausage), contained only low levels of unwanted substances. There were very rare findings of lindane and lead slightly exceeding legal maximum levels of these substances in salami.

---

[1] Schroeter A, Sommerfeld G, Klein H, Hübner D (1999) Warenkorb für das Lebensmittel-Monitoring in der Bundesrepublik Deutschland. (Market basket for food monitoring in the Federal Republic of Germany) Bundesgesundheitsblatt (Federal Health Bulletin) 1: 77-83.

- The known ubiquitous contaminants were found in nearly all **rainbow trouts** and **carps** analysed, but mostly at very low concentrations, and always below fixed maximum levels. Frequent findings of the feed additive E 324 (ethoxyquin) in trouts should be taken as an occasion to fix a maximum level.
- Benzo(a)pyrene was only found at low levels in **smoked fish** from small businesses. Only one sample exceeded the maximum level.
- 5% of cadmium contents exceeded the maximum level in **cuttle fish products**, mostly in *Sepia* species and less often in common squid. Higher concentrations were conspicuous in Asian sepia products. *Octopus* and cuttlefish in sauces or brew did not contain concentrations above the maximum level.

Cereals and cereal products

- **Rice** generally contained only low levels of undesirable substances. Maximum residue levels (MRLs) of plant protection products were not exceeded. Increased levels of arsenic and single findings of cadmium and mercury above maximum levels should be enough reason for further studies.
- **Cereal flours** dating from 2005 generally showed only low levels of the mycotoxins deoxynivalenol (DON), ochratoxin A (OTA) and zearalenone (ZEA), though the maximum level fixed for OTA was exceeded in some sparse cases in rye flours.
- **Muesli bars and mini bars** as well as **puff pastry** and **bread dough** contained only low levels of mycotoxins and heavy metals. Some concentration peaks found in muesli bars and mini bars, in particular with DON and OTA, should be further reduced or eliminated by careful selection and checks of raw materials. Levels of the reaction product 5-hydroxymethyl furfural (HMF) were in the range typical of dried fruit, which is contained in muesli bars.

Oil seed

- **Linseed** and **poppy** were slightly contaminated with OTA. Heavy metal concentrations were also low, apart from cadmium. To persistently reduce cadmium levels in oil seeds, efforts should be made to bring only such oil seeds which were grown on low-cadmium soils on the markets.

Potatoes, vegetables, mushrooms, and products therefrom

- **Potatoes, artichokes, broccoli** and **carrots** carried only low levels of residues of plant protection products, heavy metals and nitrate. 56–75% of all samples were free from measurable residues of plant protection products, and only one to two samples carried residues with one substance above the maximum residue level (MRL).
  Contents of the poisonous glycoside alkaloids solanine and chaconine in potatoes were harmless.
- Deep-frozen **spinach** carried low and fresh spinach and **French beans, cucumbers** and **tomatoes** (from conventional cropping) medium levels of plant protection product residues. 5 - 8% of residues exceeded MRLs. Residue findings in tomatoes from organic farming were not above MRLs, but nearly as frequent as findings in conventionally grown tomatoes.
- Findings of heavy metals were low, in total. Yet, cadmium levels in spinach were increased again, which is why it should be recommended to grow spinach only in low-cadmium soils. Minimisation of nitrate in fresh spinach should also be made a strategic goal, because concentrations continue to be high compared with deep-frozen spinach, sometimes exceeding the fixed maximum level.
- Special analyses of **leaf and root vegetables** for herbicide residues showed that every third substance found was a herbicide. This means herbicide findings were relatively frequent. Referred to maximum residue levels, vegetables carried medium-range levels of herbicide residues.
- No HMF and only little acrylamide was found in the ready-to-eat products **potato fritters, croquettes, potato puree powder** and **potato dumpling powder**. Most element (heavy metal) levels were low in potato puree and potato dumpling powders. Single findings of lead and cadmium near or above fixed maximum levels for potatoes should be taken as an occasion to search for potential sources of contamination (such as habitat or processing-related factors), with the aim to minimise contents.
- Related to average heavy metal levels in fresh mushrooms, findings in **dried shiitake** and **tinned champignons** were generally low. Yet, tinned champignons showed medium-range levels of tin and dried shiitake some increased cadmium findings. Concentrations of heavy metals in mushroom products and fresh mushrooms destined for processing should therefore be further surveyed. A point should be made of keeping the substrate for mushroom culture free from heavy metals as far as possible, on the one hand, and minimising contamination by processing and tin material, on the other.

Fruit, fruit juices, and other fruit products

- **Fruit, fruit juices,** and other **fruit products** contained only low levels of heavy metals.
- More than 85% of **pears, peaches** and **nectarines,** and nearly all unpeeled **oranges** and **tangerines** contained residues of plant protection products, but average levels were generally low. Regarding contents above maximum residue levels, pears and nectarines contained only low, tangerines medium-range, and oranges increased levels of plant protection product residues. Non-compliance with maximum residue levels was between 4.6 and 5%. In pears, the level of non-compliance has clearly decreased compared to 2002. About oranges and tangerines, it has to be noted that the pulp as the edible part contains only minor residue levels, as it was shown in earlier monitoring studies.
- Peaches had a high share of non-compliance with MRLs, with 15,3%. It should be made a point to improve the residue situation by suitable minimisation measures.
- As in the 1996 monitoring study, **apple juice** showed frequent findings of patulin, with slightly higher concentrations overall, and one case of MRL non-compliance. This means that apple juice producers have to take particular care that no spoilt fruit is entering the press.

Special studies looking into the use of the fungicide carbendazim produced no, or only very sparse, findings of this substance in **orange juice** and **pear juice**, but more frequent findings, though with very low concentrations, in **apple juice** and **grape juice.**

- Findings of mycotoxins in **partially fermented grape must** (such as Federweißer and young wine) and in **quality sparkling wine** were generally low, compared with fixed maximum levels. Single findings of high OTA concentrations above the maximum level should still be an occasion to pay more attention to possible mould in wine grapes.

Other foods

- **Marzipan/persipan raw matter** held only low contents of aflatoxins and heavy metals. The same holds in principle for **sweets from other raw materials.** Yet, there were more frequent findings of increased lead levels and in some cases also increased cadmium levels. The causes of these findings should be identified and eliminated. HMF levels were comparatively low.

- Furan was frequently found in **infant food** and **ready-to-eat meals,** such as soups. Though the levels found do not pose any health risk, according to what is currently understood, these levels should be further reduced.

- Lead and cadmium were frequently found in food supplements, such as **preparations of vitamins, mineral substances, plant extracts,** and **algae.** Extremely increased cadmium levels in some algal preparations were conspicuous. It seems reasonable to reduce the contaminant load to what is technologically possible and unavoidable by legal control of the maximum level. The heavy metal content of algal preparations should be further monitored in the framework of routine control action.

2 Objectives and organisation

Food monitoring aims to collect representative data on the occurrence of undesirable substances in foods in the Federal Republic of Germany and early recognise the risk potentials of these substances. In the long run, monitoring is intended to highlight trends in the contamination of foods and create a data basis sufficient to allow calculation and evaluation of the dietary intake of undesirable substances.

Food monitoring has been carried out as an independent task of official food control since 1995 on the legal basis of § 46 c–e of the Food and Commodities Act, and §§ 50–52 of the Food and Feed Code since the new food law became effective on 02 September 2002. It is an important instrument to improve preventive health protection of consumers.

Foods studied in the monitoring programmes of the years 1995 to 2002 were selected from a market basket reflecting people's consumption behaviour. The findings were the basis to determine and assess consumers' dietary intake of undesirable substances. The results were presented in a report published under the title "Results of the national food monitoring of the years 1995–2002".

A survey of all foods analysed in the framework of national monitoring schemes between 1995 and 2005 is given in chapter 7 of the present report.

Since 2003, food monitoring has been carried out as two complementary programmes. One programme included foods prescribed by an official skeleton monitoring plan for the 2005 to 2009 period developed on the basis of a representative market basket containing some 120 foods. This part is called the market basket monitoring and serves to watch the development of the situation of contamination while ensuring representative conditions of sampling.

In addition to that, special topical problems were studied in the manner of specific projects. This part of the programme is called project monitoring.

Analysis of the chosen foodstuffs was carried out by the states' (Länder) official laboratories.

Organisation of the monitoring scheme, collection and storage of data, and evaluation and reporting of monitoring findings are the tasks of the Federal Office of Consumer Protection and Food Safety (Bundesamt für Verbraucherschutz und Lebensmittelsicherheit, BVL), and here of Unit 107, a team which is dealing with food monitoring, residue control programmes, and data compilation.

What happens with the findings of food monitoring?

The findings of food monitoring are used to permanently update health risk assessments and to check and, if necessary, revise maximum levels or maximum residue levels of undesirable substances in foodstuffs. To this end, the data are provided to the Federal Institute of Risk Assessment (*Bundesinstitut für Risikobewertung, BfR*) according to § 51(5) of the Food and Feed Code. Conspicuous findings may entail further studies in future official food control and surveillance programmes.

Cases of non-compliance with legal maximum limits are pursued by the authorities of the federal states and punished if necessary. Maximum levels of residues or contaminants in and on foods are fixed in Europe and in Germany according to the principle of minimisation, that is, as low as possible under the production conditions given and with consideration of Good Agricultural Practice, but never higher than toxicologically acceptable. So, when maximum levels are established, the calculation takes account of toxicological exposure limits, such as the Acceptable Daily Intake (ADI), or the Acute Reference Dose (ARfD). These exposure limits include safety factors, mostly the factor 100, so that occasional non-compliance with maximum levels does not mean a health risk to consumers. Still, fixed maximum levels must be met by producers and distributors of a product. Products which do not comply with maximum levels must not be put on the market.

If a food is found to contain critical levels of contaminants for which permissible maximum levels have not yet been fixed, food safety authorities will make a health risk assessment, which will also consider toxicological exposure limits and consumption amounts.

In cases where dietary exposure to unwanted substances cannot be practically avoided, and where dietary recommendations do not make sense as a wide variety of foods are concerned, technological minimisation measures must be launched. This concerns substances which are formed during the food production process, such as acrylamide or furan, and for contaminants which have entered the product from the environment, such as cadmium, bromide, and nitrate. In particular, this concerns substances with mutagenic or carcinogenic effect for which a maximum level has not been fixed because any level might be harmful, or because there is no sufficient data basis for a substantiated risk assessment.

The data underlying this report are published in a tabular survey entitled "Tabellenband zum Bericht über die Monitoring-Ergebnisse des Jahres 2005" on the BVL website (German version).

The monitoring reports published so far are also available on the BVL website under http://www.bvl.bund.de, foodmonitoring.

3 The 2005 monitoring plan

The institutions of the Federal Government and of the states (*Länder*) responsible for the monitoring programme work out a detailed plan for the monitoring scheme on the basis of the General Administrative Rules on the Performance of Food Monitoring (AVV LM). This plan comprises the selection of foods and substances to be examined, as well as methods of sampling and analysis. The plan may be drawn from the "Handbuch Lebensmittel-Monitoring 2005" (Manual of the 2005 Food Monitoring), which is also accessible by internet under http://www.bvl.bund.de (German version), Lebensmittel < Sicherheit und Kontrollen < Lebensmittel-Monitoring.

As explained before, food monitoring was carried out as a dual programme. Part of the foods was again selected according to the 2005–2009 skeleton plan based on the representative market basket. That part of the programme was aimed at obtaining an overview of the situation of contamination while maintaining representative sampling conditions. The EU-wide co-ordinated Community monitoring programme checking compliance with maximum levels of pesticide residues (explanation see Glossary) is an integral part of the market basket monitoring. The co-ordinated Community monitoring programme includes only food products of vegetal origin. The other part of the programme dealt with particular problems in the form of specific projects.

3.1
Selection of foods and substances for market basket monitoring

Four foods of animal origin and 18 foods or food groups of vegetal origin were selected from the market basket to be included in the 2005 monitoring scheme. Table 3-1 lists foods and food groups and the substances analysed therein.

The range of substances to be looked for was extended again in the wake of recent findings on potential residues and contamination of foods and the introduction of further analytic methods. Namely fruit and vegetable samples examined in the framework of the market basket monitoring were analysed for up to 130 different organic substances, most of these being residues of plant protection products. Improvements of analytic measuring devices have sometimes considerably enhanced the sensitivity of analytic methods, so that very low contents and even just traces of more residues of plant protection products were found. This has allowed to produce substantiated theses about the residue situation in the foods examined in Germany.

3.2
Selection of foods and substances for project monitoring

The foods and substances or substance groups studied in separate projects in the framework of the national food monitoring scheme were selected purposeful on the basis of recent findings which indicated that specific action is needed. Table 3-2 below lists the monitoring projects.

3.3
Sampling and analysis

Sampling was performed according to the procedures described in the official collection according to § 64 Food and Feed Code (LFBG), formerly § 35 of the Foods and Commodities Act. Samples were drawn at all stages of the food chain, from producers to wholesalers and distributors to retailers.

Sampling and analysis of samples are tasks of food control authorities and official laboratories of the states. All laboratories are accredited laboratories, as required by Regulation (EC) No. 882/2004[1] on additional measures of official food control.

In order to obtain comparable results, food samples were prepared for analysis by uniform steps (for instance, washing, cleaning, peeling). Regarding the choice of analytic methods, it must be sure that the methods produce exact results and meet the validation criteria of Regulation (EC) No. 882/2004. In order to test foodstuffs for the sometimes very wide variety of organic substances planned, most of the methods used are multiple methods. Yet, some substances require single analytic methods, which raises the laboratory expense considerably. The reliability of analytic findings was checked by internal quality assurance measures, such as use of proper reference materials and participation in laboratory efficiency tests.

---

[1] Regulation (EC) No. 882/2004 of the European Parliament and of the Council of 29 April 2004 on official controls performed to ensure the verification of compliance with feed and food law, animal health and animal welfare rules. Official Journal of the European Union L 291/1 of 29 April 2004.

Table 3-1 Foods included in the market basket monitoring and substances/substance groups analysed therein in 2005.

Food	Studied in the 1995–2004 Monitoring	Substance groups/substances
Raw sausages	No	Persistent organo-chlorine compounds, nitro musk compounds, polycyclic aromatic hydrocarbons, histamin, elements
Salami	1999	Persistent organo-chlorine compounds, nitro musk compounds, histamin, elements
Carp	1997, 1998	Persistent organo-chlorine compounds, nitro musk compounds, elements
Trout	1995, 1996	Persistent organo-chlorine compounds, nitro musk compounds, elements
Rice	2000, 2003	Plant protection products, elements
Puff pastry, bread dough, muesli bars, muesli mini bars	No	Elements, mycotoxins, 5-hydroxymethylfurfural
Linseed	1999	Elements, ochratoxin A
Poppy seed	No	Elements, ochratoxin A
Potatoes	1998, 2002	Plant protection products, elements, nitrate
Potato fritters, croquettes, potato dumpling and puree powder	No	Elements, 5-hydroxymethylfurfural, acrylamide (in some samples)
Spinach	1998, 2002	Plant protection products, elements, nitrate
Artichocke	No	Plant protection products, elements, nitrate
Broccoli	1997	Plant protection products, elements, nitrate
French beans	1995, 1996, 2002	Plant protection products, elements, nitrate
Carrots	1998, 2002	Plant protection products, elements, nitrate
Champignon, tinned; Shiitake mushroom, dried	No	Elements
Pear	1998, 2002	Plant protection products, elements
Peach/nectarine	1998, 2002	Plant protection products, elements
Orange/tangerine	1996, 1998, 2002	Plant protection products, elements
Pineapple juice, apple juice, grapefruit juice	Apple juice in 1995, 1996	Elements, patulin (in apple juice only)
Quality sparkling wine, partially fermented grape must	No	Elements, mycotoxins
Marzipan raw matter, persipan raw matter, sweets from other raw materials	No	Elements, mycotoxins, 5-hydroxymethylfurfural

Table 3-2 Survey of special monitoring projects in 2005.

Foodstuffs	Specific problem	Project name
Bouillon and stock products, ready-to-eat meals, sauce powders, infant food	Furan in foodstuffs	Project 1
Juices of grapes, apples, pears, oranges and mixed juices	Carbendazim in fruit juices	Project 2
Potatoes	Glycoside alkaloids in potatoes	Project 3
Preparations of vitamins, mineral substances, plant extracts and algae	Heavy metals in food supplements	Project 4
Tomato	Residues of plant protection products in tomatoes	Project 5
Cucumber	Persistent organo-chlorine compounds and plant protection product residues in glasshouse cucumbers	Project 6
Rye and wheat flour	OTA, DON and ZEA in cereal flours	Project 7
Cuttlefish products of sepia, octopus, squid	Cadmium in cuttlefish products	Project 8
Smoked fish	Benzopyrene in smoked fish	Project 9
Basil, savoury, dill, lamb's lettuce, cress, kitchen herbs, parsley, sage, chives, spinach, thyme, lemon balm, carrot, celeriac	Herbicide residues in certain vegetables and fresh herbs	Project 10

4 Number of samples and origin

The volume of random sampling for monitoring purposes is usually fixed at 240 samples per food. This will ensure the representative character of samples and allows to make statistical statements with the certainty required.

The co-ordinated Community monitoring programme on plant protection product residues prescribes 93 samples per food in Germany. Therefore, the number of samples was reduced to about 100 with foods for which there were findings from

Table 4-1 Sample numbers (n) and origin of the foodstuffs in market basket monitoring

Origin	Domestic		EU		Third country		Unknown		Total
Foodstuffs (with food code)	n	%	n	%	n	%	n	%	n
80106 Salami	28	18.5	123	81.5					151
80300 Raw sausages	156	92.9					12	7.1	168
102615 Rainbow trout	119	97.5	2	1.6	1	0.8			122
102960 Carp	86	100.0							86
150600 Rice*	48	44.4	12	11.1	22	20.4	26	24.1	108
161113 Muesli bar/mini bar	142	92.2					12	7.8	154
161401 Wheat-flour bread dough	71	100.0							71
161505 Puff pastry	56	86.2	1	1.5			8	12.3	65
230402 Poppy seed	46	63.0	10	13.7	1	1.4	16	21.9	73
230403 Linseed	26	35.6			14	19.2	33	45.2	73
240100 Potatoes	87	85.3	6	5.9	3	2.9	6	5.9	102
240306 Potato fritter (cooked)	52	77.6	9	13.4			6	9.0	67
240308 Croquettes (cooked)	57	78.1	9	12.3			7	9.6	73
240506 Potato puree powder	60	87.0	2	2.9			7	10.1	69
250114 Spinach	125	81.7	22	14.4			6	3.9	153
250201 Broccoli	50	70.4	19	26.8			2	2.8	71
250204 Artichoke*	6	11.1	42	77.8	5	9.3	1	1.9	54
250312 French beans	64	48.9	26	19.8	15	11.5	26	19.8	131
250401 Carrots	81	77.1	22	21.0			2	1.9	105
280101 Champignon, tinned	37	45.1	32	39.0	4	4.9	9	11.0	82
280303 Shiitake, dried	38	50.7	3	4.0	28	37.3	6	8.0	75
290202 Pear	27	25.0	54	50.0	26	24.1	1	0.9	108
290303 Peach/nectarine			131	94.2	6	4.3	2	1.4	139
290401 Orange*	1	0.8	68	57.1	43	36.1	7	5.9	119
290402 Tangerine			20	100.0					20
310601 Apple juice	113	95.0	2	1.7			4	3.4	119
311601 Grapefruit juice*	53	81.5					12	18.5	65
312101 Pineapple juice*	41	80.4					10	19.6	51
334200 Quality sparkling wine	99	71.7	34	24.6	5	3.6			138
339000 Partially fermented grape must	70	93.3	5	6.7					75
431601 Marzipan raw matter	30	62.5					18	37.5	48
431900 Confectionery from other raw matters	62	80.5	10	13.0	2	2.6	3	3.9	77
Total	1931	64.1	664	22.0	175	5.8	242	8.0	3012

* With the foods marked with a starlet, the origin usually does not reflect the country of origin of the raw product but the country where the product was processed or packed.

previous monitoring programmes and which had to be examined again under the EU programme.

A total of 5,159 samples were analysed in 2005. Most of the samples were taken in retail shops, but some also at the producer's or importer's establishment. Figure 4-1 shows the shares of foodstuffs or animal and vegetal origin in the total number of samples. Bouillon and stock products, ready-to-eat meals, sauce powders, infant food, soups and food supplements were categorised as "others" in this graph. Figure 4-2 shows shares of samples of domestic and foreign origin.

Tables 4-1 and 4-2 both give a breakdown of sample numbers by origin, Table 4-1 for the market basket monitoring and Table 4-2 for project monitoring.

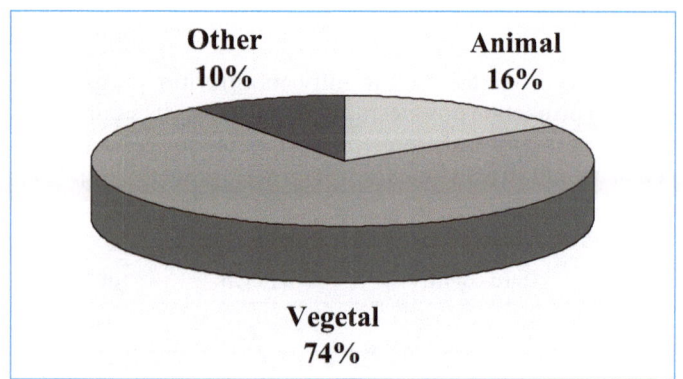

Figure 4-1 Shares of samples of animal/vegetal/other origin.

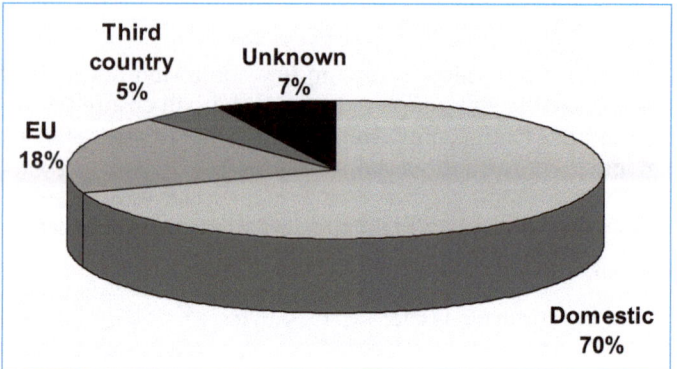

Figure 4-2 Shares of domestic/foreign origin.

Table 4-2 Numbers (n) and origins of samples in project monitoring.

Origin Project	Domestic n	%	EU n	%	Third country n	%	Unknown n	%	Total n
Furan in foodstuffs	196	95.1	5	2.4	5	2.4			206
Carbendazim in fruit juices	184	94.8	4	2.1			6	3.1	194
Glycoside alkaloids in potatoes	177	79.7	24	10.8	16	7.2	5	2.3	222
Heavy metals	264	86.3	4	1.3	4	1.3	34	11.1	306
Residues of plant protection products in tomatoes	72	33.5	135	62.8	8	3.7			215
Persistent organo-chlorine compounds in glasshouse cucumbers	204	79.7	48	18.8	2	0.8	2	0.8	256
OTA, DON and ZEA in cereal flours	242	98.4	2	0.8			2	0.8	246
Cadmium in cuttlefish products*	14	12.0	42	35.9	45	38.5	16	13.7	117
Benzo(a)pyrene in smoked fish	121	68.8	1	0.6	1	0.6	53	30.1	176
Herbicide residues in certain vegetables	178	85.2	18	8.6	4	1.9	9	4.3	209
Total	1652	76.9	283	13.2	85	4.0	127	5.9	2147

* With the foods marked with a starlet, the origin usually does not reflect the country of origin of the raw product but the country where the product was processed or packed.

5

Findings of market basket monitoring

This chapter presents the findings about the market basket foodstuffs examined under the 2005 monitoring scheme.

> All the statements of the present report regarding the situation of residues and contamination in foodstuffs solely refer to the substances and substance groups analysed in the course of the 2005 monitoring.
>
> The meaning of the criterion 'frequently quantified' depends on the substance group. With plant protection product residues and mycotoxins, frequently quantified substances were such quantified in more than 10% of samples. With organic contaminants and elements, frequently quantified substances where those quantified in more than 50% of samples.
>
> For a classification of the degree of contamination, see glossary under items 'degree of contamination' and 'nitrate'.
>
> The terms 'non-compliance with maximum residue levels' (MRLs) or 'non-compliance with maximum levels' refer to samples with contents above maximum permissible levels as fixed in legal regulations.

5.1
Sausages

Raw sausages

There is a wide variety of raw sausages on the market. They may be firm and sliceable, such as salami and cervelat, or spreadable, such as soft-pork *Teewurst* and *Mettwurst*. They are mostly made of unheated muscle and fat tissue plus spices, curing salt and sugar.

The monitoring studies included air-cured, unsmoked salami (151 samples) and various kinds of smoked *Tee-* and *Mettwurst* (168 samples). These were analysed for 25 persistent organo-chlorine compounds (including PCB congeners), histamine, nitro-musk compounds, nitrofen, polycyclic aromatic hydrocarbons (PAH, only in smoked *Tee-* and *Mettwurst*), and for six elements.

Salami was already intensively tested in the framework of the 1999 monitoring scheme, which allows comparison now.

Only 28 samples of the sliceable salami, that is 18.5%, but nearly all of the spreadable sausage samples (156; 92.9%, respectively) stemmed from German production. Salami stemmed mostly from France (32.5%) and Italy (26.5%), but also from Spain and Hungary (either to 11.9%).

Organic substances

51% of the analysed soft sausage samples and 61% of the salami samples did not contain measurable residues of persistent organo-chlorine and musk compounds. None of the organic substances looked for was found in more than 50% of samples. Substances measured to quantifiable amounts in more than 10% of samples were HCB, p,p'-DDE and lindane in salami and HCB, p,p'-DDE, beta-HCH, PCB 138 and PCB 153 in soft sausage. Persistent organochlorine compounds, namely DDT, HCB and PCB 153, were found much more frequently in salami from German production than in such from Italy or France.

Contents were usually very low, and comparable to those found in other kinds of sausage in previous monitoring studies. Mostly, concentrations were in the range of the determination limits between 0.001 and 0.002 mg/kg. Only one sample was found to contain lindane slightly above the legal residue level of 0.02 mg/kg. As a whole, concentrations of these substances in raw sausage were very low, which shows that the ban on the use of persistent organo-chlorine substances, which has been imposed in Germany and many other countries for many years now, has led to a slow but sustainable decrease in ubiquitous contamination.

Multiple residues (residues of several substances in one sample) were found in 21% of salami samples and 29% of the soft sausage samples. The maximum was six or seven substances found in some single samples of either kind of sausage.

None of the samples tested for the herbicide nitrofen was found to contain residues above the analytic limit of quantification of 0.002 mg/kg.

Of the nitro-musk compounds, there was only one finding of musk-xylene in salami and one of musk-ketone in soft raw sausage, and both at a very low level.

Histamine (see box) was found in 14.7% of soft raw sausage and 34.2% of the salami samples. Concentrations were on average 2.8 mg/kg in soft raw sausage and 21.5mg/kg in salami, with maximum concentrations of 127 mg/kg and 143 mg/kg, respectively. Medium concentrations were in the ranges known for these kinds of sausage[1]. There is no fixed maximum level for meat products. Yet, it is noted that all salami with histamine levels of more than 82 mg/kg came from Italy. Histamine concentrations in salami from France and Germany were roughly the same, and on average seven times lower than in salami from Italy (see Figure 5-1).

---

[1] http://www.was-wir-essen.de/download/Histamingehalte.pdf

Histamine in raw sausage

Biogenic amines, including histamine, are formed from amino acids in the course of protein degradation in foodstuffs, by separation of the carboxyl group of the amino acid. Many micro-organisms are able to perform this metabolism. Most of the biogenic amines are stable in heat.

Biogenic amines may be present in fermented or ripened refined products, such as cheese or raw sausage. Yet, the presence of biogenic histamines, and of histamine in the first place, is also an indicator of decay and may cause a burning feeling on the lips and in the mouth.

At high concentrations, histamine may cause nausea, dyspnoea, irritation of the skin, palpitation of the heart, and headache. These symptoms may occur after consumption of spoilt food, in particular tuna, with increased concentrations of around 1000 mg/kg.

A legally binding maximum level exists only for histamine (400 mg/kg), and only in certain fish products. It is fixed in Regulation (EC) No. 2073/2005 on microbiological criteria for foodstuffs.

Fermented raw sausage must be expected to carry up to 150 mg/kg histamine, while both the total concentration and the pattern of biogenic amines present may vary considerably.

The content of biogenic amines in raw sausage may be influenced, apart from by the microbiological quality of the raw material, by the choice of starter cultures or certain additives, for instance, nitrite curing salt.

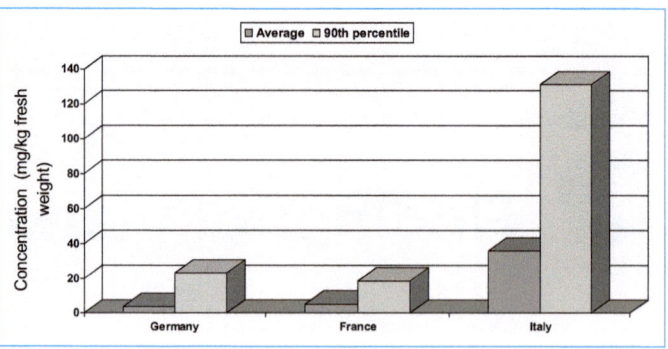

Figure 5-1 Histamine concentration in salami by origin of product.

PAH are unwanted by-products of incomplete burning processes and heating under vacuum, and may therefore be formed also in foodstuffs that are heated, dried or smoked if they come into direct contact with combustion residues. PAHs which were expected in soft raw sausage as a by-product of the smoking process were actually rarely found in the present analyses. Only benzo(a)pyrene, benzo(b)fluoranthen, benzo(k)fluoranthene and chrysene were each quantified in more than 10 per cent of the samples. Concentrations were at the analytic limit of quantification of 0.0003 mg/kg, with only very few exceptions, which were slightly higher.

Elements

Raw sausages were analysed for the elements arsenic, lead, cadmium, copper, selenium, and zinc. Table 5-1 lists the rate of findings and concentrations found for both product groups (salami and soft sausage).

Element concentrations in raw sausage were low and comparable to those found in scalded sausage in the 2004 monitoring study. The maximum values show that element concentra-

tions, in particular of copper and zinc, were generally higher in salami.

Average lead and cadmium contents in salami were nearly identical with those found in the 1999 monitoring study.

The permitted maximum level of lead of 0.1 mg/kg was slightly exceeded in five salami samples (3.3%).

Conclusion

Like other kinds of sausages examined in earlier monitoring studies, salami and soft raw sausages, including *Tee-* and *Mett-wurst*, are only slightly contaminated with undesirable substances. The concentrations found were low, apart from sparse cases of slight exceeding of legal maximum levels for lindane and lead in salami.

Histamine levels were in the known and harmless range, but levels in air-cured salami from Italy were seven times higher than in salami from Germany and France.

5.2
Fish

Rainbow trout

Of the freshwater fish, rainbow trout is a very popular, low-fat food. It is mostly produced at fish farms under controlled habitat conditions. Because of the important consumption amounts, rainbow trout was again included in the monitoring scheme in 2005, after it had already been intensively studied in 1995 and 1996. In 2005, 122 samples, most of them of domestic

Table 5-1 Element concentration in raw sausage (in mg/kg fresh weight).

Element	Share of samples with quantifiable contents (%)		Average value		Maximum value	
	Salami	Soft raw sausage	Salami	Soft raw sausage	Salami	Soft raw sausage
Arsenic	16.4	10.6	0.016	0.012	0.055	0.024
Lead	33.1	19.6	0.023	0.014	0.168	0.072
Cadmium	32.5	13.3	0.003	0.003	0.016	0.009
Copper	98.0	86.7	1.58	0.675	52.7	1.85
Selenium	95.4	86.7	0.175	0.089	1.01	0.255
Zinc	99.3	100.0	36.9	19.0	73.0	37.1

production, were again analysed for 28 persistent organo-chlorine compounds (including PCB congeners), two nitro-musk compounds, seven elements, and (new) for ethoxyquin and pendimethalin. This yielded a good data basis for a comparative consideration of the contamination found now and in earlier studies.

Organic substances

Ubiquitous persistent organo-chlorine compounds were found in nearly every sample (see Figure 5-2). Only two samples (1.6%) did not carry measurable concentrations. The fact that there were more samples with detectable residues in 2005 than in 1995 and 1996 is partly attributable to improved performance of the analytic equipment and to the wider range of substances looked for in 2005.

Concentrations were very low and mostly in the range of the analytic limit of quantification of 0.001 mg/kg. But nearly all samples (91%) carried multiple residues because of the omnipresence of these substances in the aquatic environment. Sixty-nine samples carried more than 9 substances, the maximum was 21 substances in one sample.

More than 50% of samples carried chlordane, DDT, dieldrin, HCB, PCB 52, PCB 101, PCB 118, PCB 138, PCB 153, PCB 180 und polychlorterpene (toxaphen).

Bromocyclene and musk-xylene were found much less frequently than in the earlier monitorings, with 2.5% and 19.7%, respectively. (For comparison: 37.2% and 57.4% in the 1995 and 18.8% and 43.6% in the 1996 monitoring schemes.) There were no findings above the maximum permissible levels, concentrations reaching a maximum 0.001 mg/kg.

The herbicide pendimethalin was quantified in 4.8% of the samples only, and to a maximum of 0.008 mg/kg. The source of this substance in fish is not yet clear. Given the very low concentrations, a possible source might be run-off of herbicide from agricultural fields into field-bordering water bodies.

Ethoxyquin, in contrast, was found in 69% of all samples. It is an anti-oxidant agent which is authorised as a feed additive (E324) with up to 150 mg/kg feed. Measured concentrations were 0.007 mg/kg on average, and a maximum 0.08 mg/kg. Ethoxyquin findings of that kind were evaluated in general by the Federal Institute for Risk Assessment (*Bundesinstitut für Risikobewertung, BfR*), which came to the conclusion that a person's acceptable daily intake (ADI) of 0.005 mg/kg body weight is covered only to small degree (about 8%) by such amounts of ethoxyquin in foods[2].

Elements

Rainbow trout was analysed for the elements arsenic, lead, cadmium, copper, mercury, selenium and zinc. Apart from lead and cadmium, these elements were found frequently or even in every sample. Lead was quantified in 15.6%, and cadmium only in 6.6% of the samples.

Average levels were low, and maximum levels were not exceeded. Figure 5-3 shows the concentrations of lead, cadmium, and mercury. The levels are too low to allow trend statements.

Conclusion

Nearly all rainbow trout produced in fish farms carry several of the known ubiquitous environmental contaminants at once, but with very low concentrations. The degree of contamination is largely comparable with that found in the years 1995 and 1996.

Frequent findings of the antioxidant agent ethoxyquin are conspicuous. Obviously, this substance enters the fish with the feed, because it is authorised as feed additive E 324. Here it is necessary to regulate the contents with a maximum permissible level in farmed fish.

Carp

Carp is also highly valued as a fish for human consumption, and most carp on the market are produced on fish farms. It was intensively studied in the framework of the 1997 and 1998 food monitoring schemes. Study results of the year 2005 now allow to draw conclusions about the trend of contamination with unwanted substances.

Eighty-six carp samples were taken, all from domestic production, and analysed for 28 persistent organo-chlorine compounds (including PCB congeners), two nitro-musk compounds, seven elements, and residues of ethoxyquin and pendimethalin.

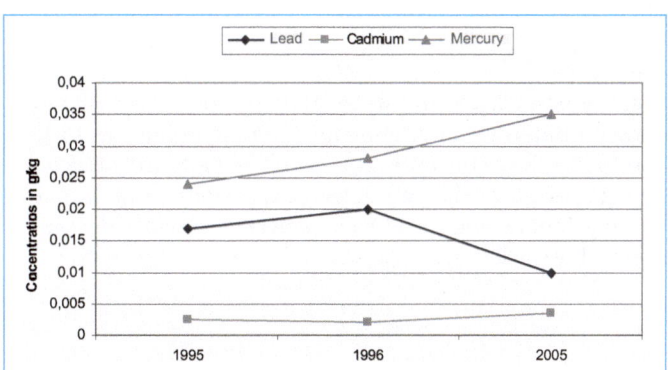

Figure 5-3 Average concentrations of heavy metals in trout in a comparison of years. (Permissible maximum levels are 0.2 mg lead/kg, 0.05 mg cadmium/kg and 0.5 mg mercury/kg.)

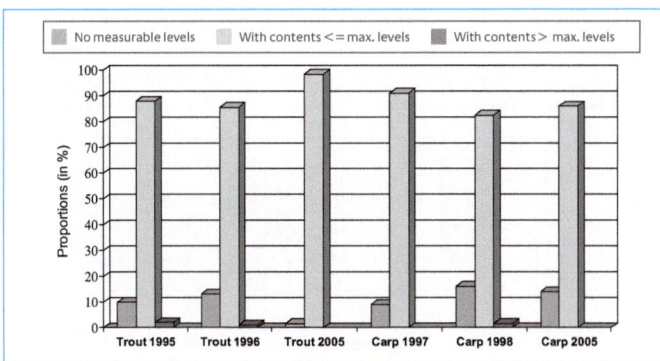

Figure 5-2 Proportions of samples of trout and carp carrying organo-chlorine, compared in different years.

---

[2] http://www.bfr.bund.de/cm/208/bewertung_der_ergebnisse_des_nationalen_rueckstandskontrollplans_2005.pdf

Organic substances

Similar to trout, the ubiquitous persistent organo-chlorine compounds were also found in 86% of the carp samples. Obviously, there has been hardly any change about the presence of these substances in the aquatic environment over the past few years, the portion of positive findings remaining nearly the same as in 1998 (see Figure 5-2).

Yet, actual concentrations were very low and under 0.01 mg/kg, apart from a few exceptions, and thus under the legal maximum residue levels. Some substances, namely lindane and PCB congeners, showed a clear decline compared to 1998.

The share of samples carrying multiple residues was 59%. The majority of this portion of samples carried 6 residues at a time, three samples carried even ten residues.

As in the 1998 monitoring, DDT was the only substance found really frequently. HCB, lindane, and the PCBs 101, 138, 153, and 180 were each found in more than 20% of samples. Bromocyclene and musk-ketone were found in only one sample each (that is 1.2%), and at a very low concentration. This is a clear decline in frequency compared to 1998, when bromocyclene was found in 6.2% and musk-ketone in 10.9% of samples.

In contrast to findings in rainbow trout, ethoxyquin was found only once in carp, with a concentration of 0.008 mg/kg. This study cannot answer the question whether carp farmers use different feed, which does not contain ethoxyquin, or whether this feed additive is just not accumulated in this fish to the same degree as in trout.

Pendimethalin was found in 15.8% of carp samples, that is, more frequently than in trout. The maximum concentration found was 0.009 mg/kg.

Elements

Carp samples were also analysed for the elements arsenic, lead, cadmium, copper, mercury, selenium, and zinc. As in trout, most elements were found frequently or even in all samples, apart from lead and cadmium, which were detected only in 26.7% and 11.6% of samples, respectively.

Average concentrations were comparable with those of 1998 and at low levels, similar to those found in rainbow trout. There were no concentrations above legal maximum levels.

Conclusion

Even though carp from fish farms often carried several of the known ubiquitous environmental contaminants – as in 1997 and 1998 – their concentrations were very low and always below the permitted maximum levels. With some contaminants, namely lindane and the PCB congeners, levels have decreased noticeably since 1998.

5.3
Cereals

Rice

Rice has already been examined for undesirable substances in the years 2000 and 2003 because of its great importance in human nutrition. Contamination was found generally low at that time. In 2005, rice was again included in the monitoring scheme because of the EU Commission's recommendations for the co-ordinated Community monitoring programme.

A total of 105 rice samples were examined for residues of 132 plant protection products and seven elements.

Plant protection products

Compared to findings in the year 2000, the share of samples containing quantifiable residues has increased to 45% in 2005 (see Figure 5-4). The share of samples with multiple residues was 21%, with a maximum of 5 substances found in one sample.

The increase in positive findings is attributed to improved analytic technology compared to earlier investigations, and to the fact that many more substances are being looked for today. Of the 132 substances looked for, only 18 were found, and these mostly at less than 0.01 mg/kg. Among these 18, only two substances appeared frequently – piperonyl butoxide, which is used as a synergist to pyrethrum, and bromide, as in the year 2000. Bromide being a ubiquitous substance, it can be assumed that most of the bromide findings are of natural origin and must not necessarily be attribute to the use of bromocontaining fumigants. There was no non-compliance with maximum residue levels.

As a whole, the findings confirmed findings of earlier monitoring studies, according to which rice carries only very low residues of plant protection products.

Elements

Rice samples were analysed for arsenic, lead, cadmium, copper, mercury, selenium, and zinc. Lead and mercury were found in only 25.0% and 19.4% of the samples, respectively, while the other elements were found frequently or even in all samples.

Element levels were roughly similar to those in the year 2000, and in particular to those of 2003. Concentrations of lead, cadmium, and mercury were mostly below 0.1 mg/kg. As in earlier monitoring studies, levels of arsenic were found increased, with an average concentration of 0.14 mg/kg and a 90th percentile concentration of 0.27 mg/kg in 2005. There is no maximum level established for arsenic.

Three samples (3.1%) exceeded the maximum permissible level of mercury (0.01 mg/kg), and one sample (0.9%) the maximum level of cadmium. Copper levels were all below the German legal maximum level of 10 mg/kg.

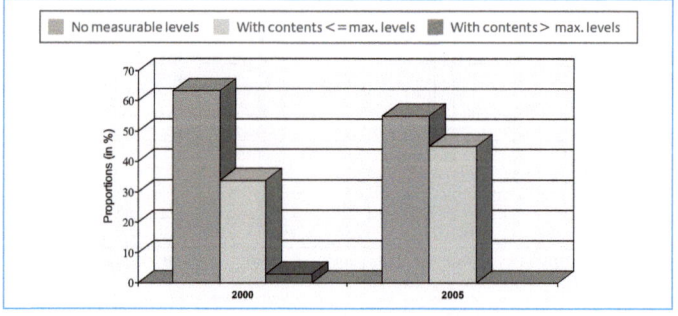

Figure 5-4 Comparison of plant protection product residues in rice.

Conclusion

The monitoring studies of the year 2005 have confirmed the findings of the 2000 and 2003 studies saying that rice is hardly contaminated with the substances in question. Yet, relatively high levels of arsenic and contamination with cadmium and mercury should continue to be monitored, because it cannot safely be precluded that legal maximum levels are exceeded.

5.4
Cereal products

Muesli bars, muesli mini bars

Muesli bars or mini bars are sold as a wide variety of energetic snacks for sports and leisure time. They are very popular as fast intermediate food or as a sweet to go with a bigger meal. Like muesli, they consist of different kinds of cereal flakes and flours, nuts, oil seed, dried fruit, milk portions, cocoa, chocolate, sugar, aroma agents and so on.

As the different components of muesli bars may carry different kinds and degrees of contamination, 154 muesli bar samples were examined for mycotoxins, elements and for HMF (5-hydroxymethylfurfural). HMF is a reaction product of sugar which is formed during heating processes or improper storage.

More than 90% of the samples stemmed from German production.

Mycotoxins

Muesli bar samples were tested for the aflatoxins B1, B2, G1, G2, deoxynivalenol (DON) and ochratoxin A (OTA). Aflatoxin B1 was found in 15.3% of samples, DON in 11.2% and OTA in 21% of samples, that is, relatively frequently. There was no non-compliance with maximum levels.

Contamination with DON and OTA was found more rarely and at lower levels than in breakfast cereals of similar composition, which had been studied in the food monitoring of 2004 (see Figures 5-5 and 5-6). Maximum concentrations, however, were similar to those in the breakfast cereals, with 318 µg DON/kg and 2.94 µg OTA/kg, which is in the range of the legal maximum levels of 350 µg DON/kg and 3 µg OTA/kg, respectively.

The aflatoxins B2, G1 and G2 were each found only in one sample, and at very low contents of less than 0.1 µg/kg.

HMF

HMF was found in 88% of all samples. The average content was 42 mg/kg, and the 90th percentile content 152 mg/kg, that is the range typical of dried fruit[3], which forms part of fruit-muesli preparations and muesli bars.

Elements

Muesli bar samples were analysed for arsenic, lead, cadmium, copper, nickel, mercury, selenium, and zinc. Mercury was found in very low concentration, and only in 3.5% of samples. Average contents of arsenic and lead were also very low, but both elements were quantified in 30 or 32% of the samples, respectively. All other elements were found frequently or in nearly all samples. Yet, average concentrations were also low. Only one sample slightly exceeded the legal maximum level of cadmium of 0.1 mg/kg.

Conclusion

Muesli bars and mini bars analysed in the 2005 food monitoring were only slightly contaminated with mycotoxins and heavy metals. To further reduce or eliminate rare peaks of contamination, which were observed with mycotoxins, it is recommended to concentrate on carefully selecting and checking raw materials. HMF contents were typical for dried fruit, which forms part of muesli bars and mini bars.

Puff pastry, bread dough

The 2005 food monitoring for the first time looked into possible contamination of readily prepared puff pastry and bread dough (including pre-baked, not including non-prebaked bread) with mycotoxins and heavy metals. While puff pastry is largely made of wheat flour, bread dough may contain a wide variety of ingredients, beside wheat, rye, or spelt flour. All 71 samples of bread dough and nearly all 65 puff pastry samples stemmed from German production.

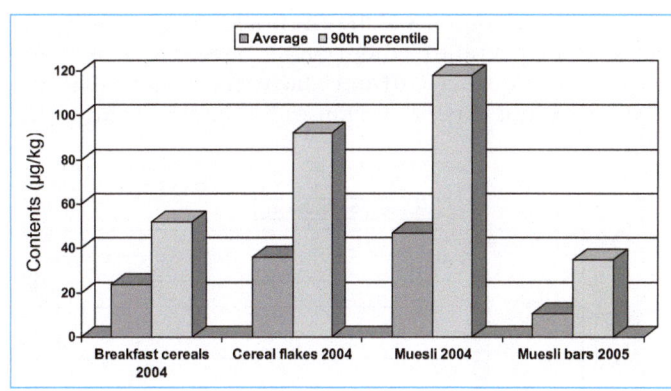

Figure 5-5 Deoxynivalenol (DON) in cereal products (for comparison: the permitted maximum level is 350 µg/kg.).

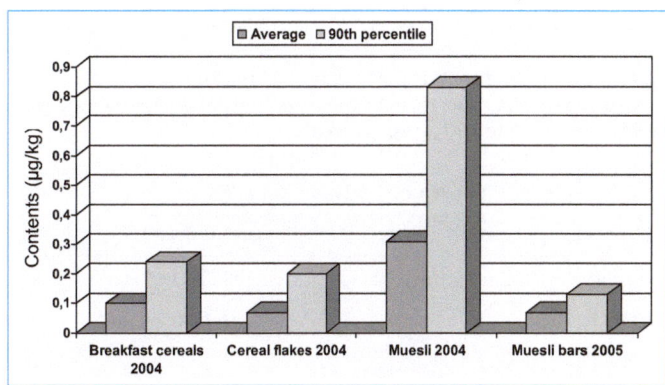

Figure 5-6 Ochratoxin A (OTA) in cereal products (for comparison: the permitted maximum level is 3 µg/kg.).

---

[3] Murkovich M. and Pichler N. (2006): Analysis of 5-hydroxymethylfurfural in coffee, dried fruits and urine. Mol. Nutr. Food Res. 50: 842–846.

Mycotoxins

DON was detected in a quarter of puff pastry samples and nearly all bread dough samples. But 90 per cent of all concentrations were lower than 110 µg/kg, and thus much lower than the legal maximum level of 350 µg/kg.

OTA was also found in a quarter of puff pastry samples and in 41% of the bread dough samples. Ninety per cent of the concentrations were below 0.6 µg/kg and thus also far below the legal maximum level of 3 µg/kg.

Figures 5-7 and 5-8 show that the contamination of bread dough with DON and OTA is roughly of the same degree as the contamination of rye from the previous year's harvest. This prompts the assumption that many of the bread dough samples were made from rye flours of the year 2004. In contrast to that, puff pastry, which is made from wheat flour, does not show a correlation with the degree of contamination of wheat of the year 2003, with DON concentrations being much lower and OTA concentrations markedly higher than those in wheat of the year 2003.

Elements

Puff pastry and bread dough samples were analysed for contamination with arsenic, lead, cadmium, copper, nickel, selenium, and zinc. Table 5-2 gives a survey over the frequency of detection and average contents of elements.

Apart from contents of the essential trace elements copper and zinc, average element contents were very low, lying in the range between 0.01 and 0.06 mg/kg. Legal maximum levels were not exceeded. Figures 5-9 and 5-10 show lead

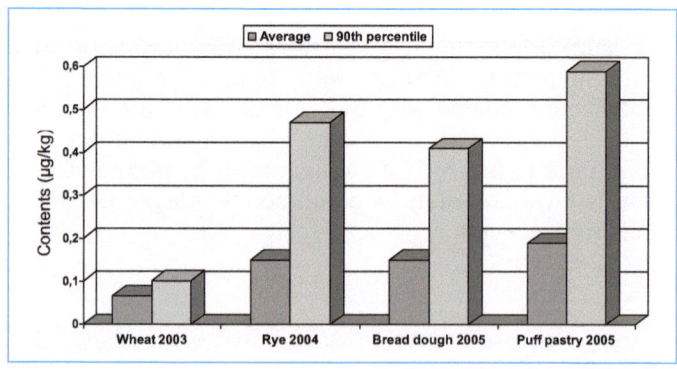

Figure 5-8 OTA contents in puff pastry and bread dough compared to cereals. (For comparison: the maximum permissible level is 3 µg/kg in cereal products.).

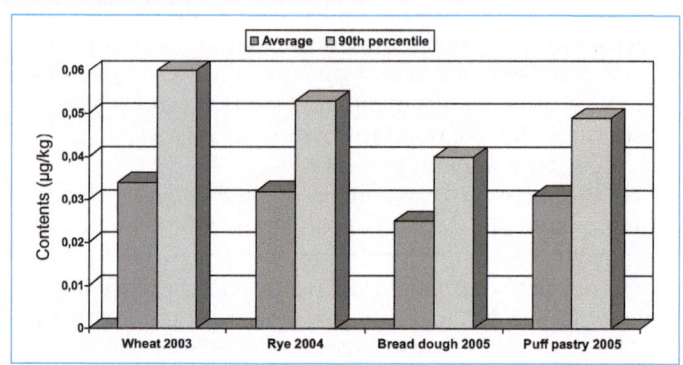

Figure 5-9 Lead contents in bread dough and puff pastry compared with cereals. (For comparison: the legal maximum level is 0.2 mg/kg in cereals.).

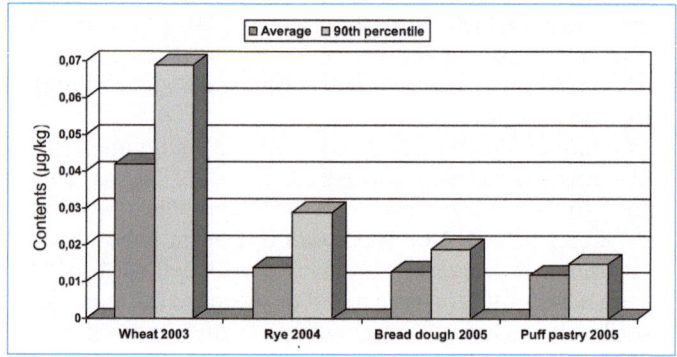

Figure 5-10 Cadmium contents in puff pastry and bread dough compared with cereals. (For comparison: the legal maximum level is 0.1 mg/kg in rye and 0.2 mg/kg in wheat.).

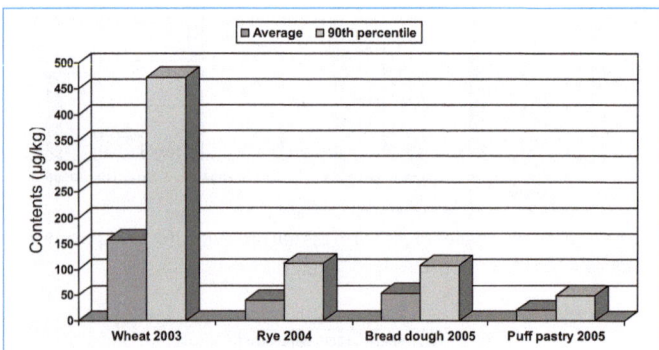

Figure 5-7 DON contents in puff pastry and bread dough compared to cereals. (For comparison: the maximum permissible level is 350 µg/kg in cereal products and 500 µg/kg in cereal grain.).

Element	Share of positive findings (%)		Average contents (mg/kg)	
	Puff pastry	**Bread dough**	**Puff pastry**	**Bread dough**
Arsenic	29.6	18.5	0.040	0.016
Lead	16.9	20.0	0.031	0.025
Cadmium	83.1	75.4	0.012	0.013
Copper	97.2	67.7	0.670	1.376
Nickel	49.3	16.9	0.066	0.113
Selenium	46.5	7.7	0.019	0.021
Zinc	100.0	100.0	3.721	8.447

Table 5-2 Shares of positive findings and average contents of elements.

and cadmium concentrations as examples. Lead levels in bread dough and puff pastry are nearly the same as in cereals. As with DON and OTA, cadmium levels were nearly the same as in rye in 2004, but markedly lower than in wheat in 2003.

Conclusion

Puff pastry and bread dough were only slightly contaminated with mycotoxins and heavy metals, confirming the positive image of cereals of the previous years.

5.5
Oil seed

Linseed, poppy

Linseed is a well-tried product maintaining the health of the gastro-intestinal tract, with a mildly laxative effect without disturbing side-effects. It is therefore also added to many foodstuffs such as bread and muesli. To what degree the linseed ingredients (including contamination of the seed) are resorbed, depends on the degree of grinding and is therefore highest with the crushed grain, and lowest with the whole grain.

Linseed was analysed for residues of plant protection products and contamination with the heavy metals lead and cadmium during the 1999 food monitoring, and found to be highly contaminated with cadmium. Apart from linseed, cocoa, poppy, and sunflower kernels count among the foodstuffs with the potentially highest cadmium loads, because these plants selectively draw cadmium from the soil and accumulate it in the seed.

Poppy was used as a nourishing food as early as in the Antiquity. For its nutty and slightly bitter aroma, the ground seed of garden poppy is a much used ingredient to sweet dishes and cakes, and the unground seed is used as a spice in bread, rolls, and cheese crackers.

To check the situation of contamination, 73 samples of brown linseed were included again and 73 poppy samples included for the first time in the 2005 National Food Monitoring, with analyses being performed for heavy metals and the mycotoxin OTA. Linseed was not analysed again for plant protection products as practically no residues had been found in 1999.

Ochratoxin A (OTA)
OTA was only found in 7.9% of the linseed samples and 14.7% of the poppy samples. Concentrations were very low and usually under 0.41 µg/kg in poppy and 0.15 µg/kg in linseed (see Figure 5-11).

Elements
Linseed and poppy were analysed for contents in arsenic, lead, cadmium, copper, selenium and zinc. Arsenic was found in a quarter of the linseed and 58% of poppy samples. Lead was found in 30% of linseed and 38% of poppy samples, and selenium in both foods in three quarters of the samples. Cadmium, copper and zinc were detected in nearly all samples as it was to be expected.

Figure 5-12 shows the cadmium concentrations, including a comparison with sunflower kernels and earlier findings in linseed in 1999. Cadmium concentrations in linseed were insignificantly lower than in 1999. Concentrations in poppy and in sunflower kernels as in 2000 were of comparable range. Maximum concentrations were 1.3 mg/kg in poppy and 0.7 mg/kg in linseed, compared with 1.1 mg/kg in sunflower kernels in 2000.

Contents of other elements were low. Average lead levels of 0.04 - 0.06 mg/kg are similar to those of other kinds of oil seed examined before. The findings with linseed were roughly the same as in 1999.

Conclusion
Linseed and poppy were only slightly contaminated with ochratoxin A, both with regard to rate of contaminated samples and concentrations found. Contamination with heavy metals is also very low, except for cadmium. Cadmium is selectively taken up from the soil by these oilseed plants, and is accumulated in the seed. This results in a relatively high concentration of cadmium in the seed. Sustainable reduction of cadmium concentrations will only be possible when these crops are grown on cadmium-poor soils.

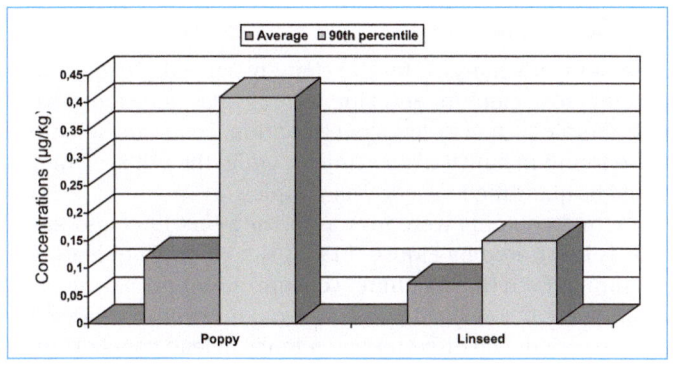

Figure 5-11 OTA concentrations in linseed and poppy.

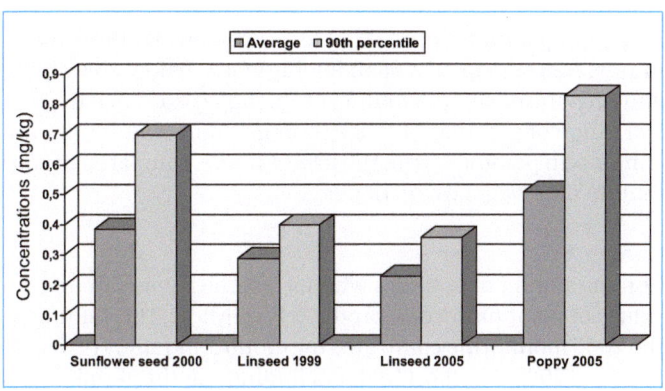

Figure 5-12 Cadmium concentrations in various oil seed compared by years.

5.6
Potatoes

Potatoes are of outstanding importance for nutrition in this country, as are cereals. They were studied in national food monitoring schemes before, in the years 1998 and 2002. Both studies showed that contamination of potatoes with unwanted substances was agreeably low. For a new check of the situation of contamination in potatoes, 102 potato samples were analysed for residues of plant protection products, elements and nitrate in the framework of the co-ordinated Community monitoring programme in 2005.

Plant protection products
Although the number of plant protection product residues looked for was much larger than in 1998 and 2002, near to three quarters of all samples were again without detectable residues. Residues of only 20 active substances were found, out of 130 substances looked for. The most frequent finding was chlorpropham with 16.7%. Chlorpropham is used as a germ inhibitor during storage and is in fact removed by washing and peeling. Pencycuron, which is used in seed potatoes before planting as a prevention against fungal diseases, was also frequently found, with a rate of 10.6%. The fungicide procymidon and dithiocarbamate fungicides exceeded maximum residue levels (MRL) in one sample each. This rate of non-compliance with MRLs of 2% is comparatively low.

The rate of multiple residues of 10% was also relatively low. The maximum was four substances found in one sample.

Elements
Potatoes were analysed for the elements arsenic, lead, cadmium, copper, selenium and zinc. Arsenic was detected only in one sample at a very low concentration. Lead and selenium were found in a fifth of all samples, while the other elements could be quantified in nearly all samples.

Actual contents were low, as in the years 1998 and 2002. This is illustrated by Figure 5-13 on the example of lead and cadmium. Even the maximum concentrations of lead and cadmium were below 0.1 mg/kg, and legal maximum levels were not exceeded. The discernible tendency in lead concentration is slightly declining over the years, while cadmium concentrations have slightly increased.

Nitrate
The average nitrate content of 130 mg/kg is nearly the same as in 1998 (128 mg/kg). 95% of all findings were below 268 mg/kg, and even the maximum finding of 391 mg/kg exceeded the typical range of potatoes of <30–350 mg/kg only slightly. Nitrate contents in potatoes are in the lowest range, compared to contents in most vegetable varieties.

Conclusion
Contamination of potatoes with nitrate, heavy metals and residues of plant protection products is only low. This finding of the 2005 monitoring confirmed the monitoring study results of 1998 and 2002. The most frequent finding was the germination inhibitor chlorpropham. This substance is actually removed by washing and peeling.

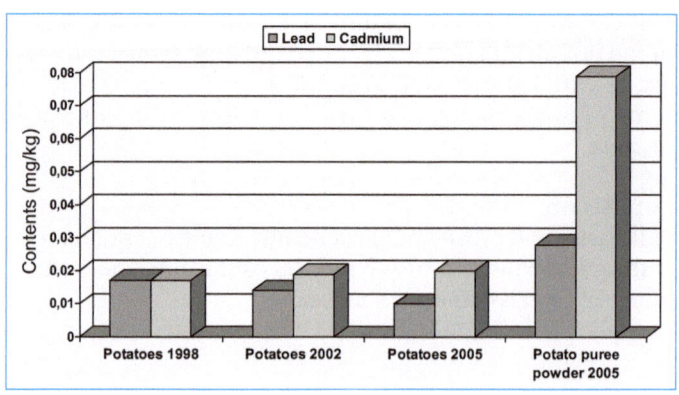

Figure 5-13 Average lead and cadmium levels in potatoes and potato products. (For comparison: the permissible maximum levels are 0.1 mg/kg in peeled potatoes.).

5.7
Potato products

Potato fritters, croquettes, potato puree powder, potato dumpling powder

The variety of finished potato products on the market is huge, and the 2005 food monitoring for the first time included four such product in the analytic scheme. Analyses included 67 samples of cooked potato fritters, 73 samples of croquettes, and 69 samples of potato puree and potato dumpling powders. More than 77% of the samples stemmed from German production. Samples were analysed for the potential heat reaction products acrylamide (only in fritters and croquettes) and HMF, and the potato puree powder and dumpling powders were also analysed for seven elements.

Acrylamide
As it had been expected, one third of the potato fritters and two thirds of the croquettes contained very low amounts of acrylamide before deep-frying. The maximum concentration found was 61 µg/kg, which is extremely less than the acrylamide concentrations of up to 3000 µg/kg found in official food control tests of readily prepared (deep-fried) fritters in the year 2005[4].

HMF (5-hydroxymethylfurfural)
HMF could not be quantified in any of these potato products.

Elements
Potato puree powder and potato dumpling powder were analysed for arsenic, lead, cadmium, copper, nickel, selenium and zinc. Arsenic, lead, and selenium were found in less than 15% of samples and at very low concentrations. Nickel was quantified in somewhat more than half of the samples, with 90% of the concentrations measured being <0.22 mg/kg and an average concentration of 0.12 mg/kg. The other elements were present in nearly all samples.

---

[4] See at http://www.bvl.bund.de/, Lebensmittel < Unerwünschte Stoffe & Organismen < Acrylamid < Kartoffelpuffer

The dehydration process alone, which forms part of the production process of these powders, already leads to an enrichment of elements. This is illustrated by a comparison with fresh potatoes in Figure 5-13. While there are no legal maximum levels for heavy metals in ready-to-eat potato products, there are maximum levels for lead and cadmium in fresh, peeled potatoes. These are 0.1 mg/kg for either element. Proceeding from these levels, 8.7% of samples would have exceeded the maximum permissible cadmium level and 2.9% of samples the maximum permissible lead level. One the other hand, if one takes account of the fact that the products do not only consist in potato, and that the water portion has practically been extracted, none of the samples would exceed the maximum level for lead, but still two of the samples (ca. 3%) would exceed the maximum level for cadmium after extrapolation of the measured concentrations to fresh potato using estimated process-related conversion factors.

Conclusion

The finished products potato fritters, croquettes, potato puree powder, and potato dumpling powder are not, or only slightly contaminated with the reaction products HMF and acrylamide. Element contents quantified in potato puree and potato dumpling powders are generally low. Some single samples with lead and cadmium levels near or above the maximum levels fixed for fresh potatoes should be taken as an occasion to spot possible sources of contamination (namely, crop habitat factors, processing factors) and try to minimise element contents.

5.8
Leaf vegetables

Spinach

Spinach was also examined in the 1998 and 2002 food monitoring schemes, and was again included in the study in 2005 in the framework of the co-ordinated Community control programme. As study results of three years can now be compared, statements about trends and differences between fresh and deep-frozen spinach should be possible. In 2005, a total of 153 spinach samples – 89 fresh and 64 deep-frozen – were analysed for plant protection products, elements and nitrate. 82% of samples were of German origin, 7% from Italy, and 11% from other countries.

Plant protection products
Although more substances were searched for in 2005, the share of fresh spinach samples without residues was larger than in 2002, and the share of residue-free samples of deep-frozen spinach was about the same as in 1998 and 2002 (see Figure 5-14). However, the rate of residue-free samples is much higher in deep-frozen spinach anyhow. That was also the case in 1998 and 2002. That is attributable to the preparation of deep-frozen spinach, including washing and blanching, but also to pesticide minimisation strategies pursued by some growers, who have concluded growing contracts or have re-introduced mechanical or biological measures of pest control. Yet, there was a higher rate of non-compliance with maximum residue levels

both in deep-frozen and in fresh spinach, with 4.7% and 5.6% of the samples, respectively.

Of 130 substances looked for, residues of 40 substances were found, most frequently of phenmedipham (26.2% of samples) and lambda-cyhalothrin (11.8% of samples). Only very few substances reached average concentrations of slightly higher than 0.01 mg/kg.

Single cases of non-compliance with maximum residue levels were found with the following substances: fungicidal dithiocarbamates (three cases) as well as endosulfan, methamidophos, tolclofos-methyl, dimethoate, chlorpropham and iprodion (one case each).

The positive effects of pesticide minimisation strategies and of cleaning during production of deep-frozen spinach are also reflected in the rate of multiple residues: only one sample of deep-frozen spinach carried two residues (maximum), while the rate of multiple residues in fresh spinach was 15.7%, with a maximum of three residues in one sample.

Elements
Spinach was analysed for contents of arsenic, lead, cadmium, copper, selenium, thallium and zinc. Selenium and thallium were found in only a quarter of all samples, while arsenic was present in 44% of the samples. Lead was detected even more frequently, that is, in 80% of the samples, and the other elements were quantifiable in all or nearly all samples.

Apart from cadmium, the levels of all elements were low. With arsenic, lead, and selenium, 90th percentile values were below 0.1 mg/kg. Thallium levels were extremely low and even at maximum only slightly higher than the limit of quantification of 0.02 mg/kg. Concentrations of copper and selenium were found slightly higher than in the previous years. Copper levels (90th percentile of 1.9 mg/kg) prompted the assumption that part of the residues stemmed from application of copper as fungicide.

Contamination with cadmium was rated as of medium degree, the same as in 1998 and 2002. In the 2005 monitoring scheme, the 90th percentile concentration was 0.16 mg/kg, and 2.7% of the samples exceeded the maximum level for cadmium of 0.2 mg/kg.

Lead reached a concentration of 0.33 mg/kg in one sample, exceeding the maximum level of 0.3 mg/kg.

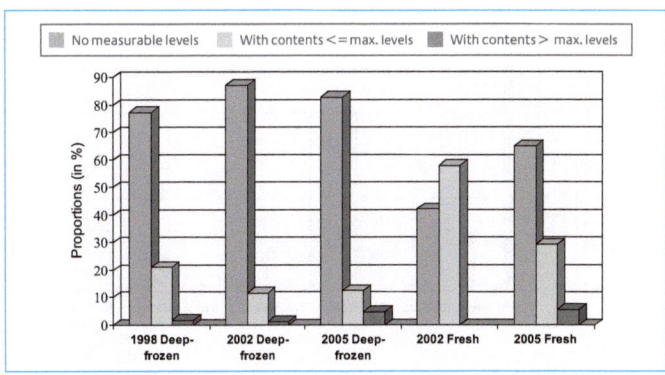

Figure 5-14 Residues of plant protection products in spinach in a comparison of years.

Nitrate

Spinach is known to carry relatively high levels of nitrate. Similar to the 2002 monitoring scheme, studies in 2005 also showed a rate of non-compliance with the fixed maximum level of 9.2% (14 samples), though only in fresh spinach this time. On the other hand, overall levels in fresh spinach have declined since 2002, as Figure 5-15 shows. Nitrate levels in deep-frozen spinach have been nearly constant since 1998 and generally clearly lower than in fresh spinach. This is attributed to producers' nitrate minimisation strategies, and to the processing steps of washing and blanching before deep-freezing.

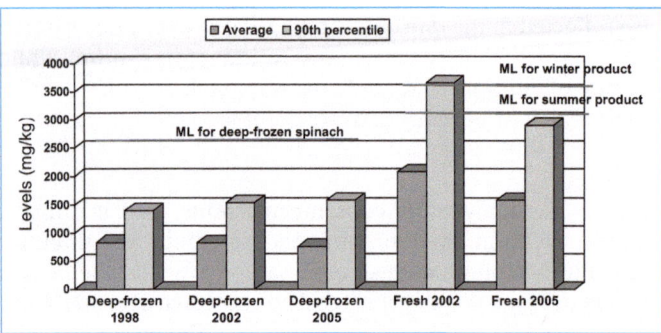

Figure 5-15 Nitrate levels in fresh and deep-frozen spinach with a comparison of years (ML: Maximum level).

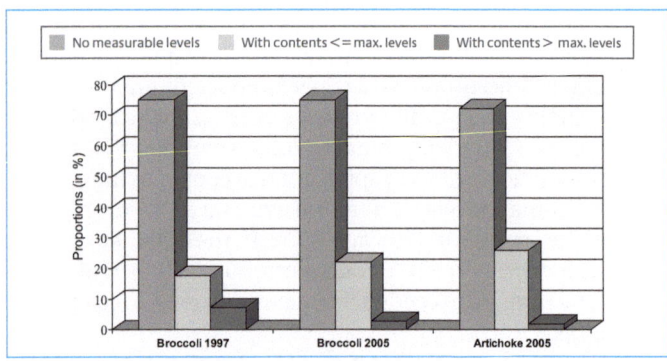

Figure 5-16 Plant protection products in artichoke and broccoli compared by years.

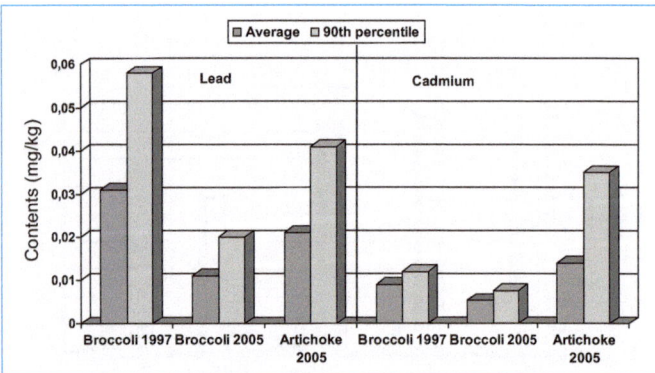

Figure 5-17 Lead and cadmium contents in artichokes and broccoli compared by years. (For comparison: the permissible maximum levels are 0.1 mg/kg.).

Conclusion

Contamination of deep-frozen spinach with plant protection product residues is low as a result of processing steps and growers' pesticide minimisation strategies, while fresh spinach is to medium degree contaminated. Contamination with heavy metals is overall low. Yet, cadmium levels were found increased, as in previous monitoring studies. It is recommended that spinach should be grown only on cadmium-poor soils. It is further recommended to develop strategies to minimise nitrate levels in fresh spinach, as concentrations continue to be relatively high here compared to deep-frozen spinach, and the maximum level was again exceeded.

5.9
Sprouting vegetables

Artichoke

Artichokes are an increasingly popular vegetable and now on the market throughout the year. They were first included in the food monitoring scheme in 2005, with 54 samples being tested for residues of 131 plant protection products, for elements and nitrate. One third of samples each stemmed from France and Italy, 9% each from Spain and Egypt, and the rest from other countries.

Plant protection products

Artichokes were only very little contaminated with plant protection products residues, as Figure 5-16 shows. The majority of samples (72%) were without measurable residues. Only one sample (1.9%) exceeded the maximum residue level of 0.02 mg/kg of procymidone, a fungicide.

Residues of only 13 substances were found, out of 131 substances looked for. Often this was only the fungicide myclobutanil (11% of all samples). Average concentrations were very low and only in one case slightly higher than 0.01 mg/kg. The proportion of samples carrying multiple residues was also relatively small with 9.3%. The maximum was four residues in one sample.

Elements

Artichokes were tested for arsenic, lead, cadmium, copper, selenium, thallium and zinc. Selenium and thallium could be quantified only in less than 4% of the samples, arsenic in 11% and lead in 24% of the samples. Cadmium was found in three quarters of the samples, and copper and selenium in nearly all. Overall element contents were low, which is shown in Figure 5-17 on the examples of lead and cadmium. There was no non-compliance with fixed maximum levels.

Nitrate

Artichokes contained only relatively little nitrate. 90% of the samples carried less than 72 mg/kg, and the maximum of 325 mg/kg was also comparatively low.

Conclusion

Artichokes are only very lightly contaminated with plant protection product residues, elements and nitrate.

Broccoli

Broccoli is a kind of predecessor of cauliflower. It is a much consumed vegetable, also because it is easily digestible compared to other kinds of cabbage and therefore suitable as dietetic food, and because it is consumable both as raw or cooked food.

Broccoli was studied before in the 1997 food monitoring. Those studies showed light contamination with lead and cadmium, and medium-degree nitrate contents, but frequent residues of plant protection products. Prevailing were traces of bromide and dithiocarbamates (DTC), which could, however, also have been of natural source. This makes a competent evaluation of the situation difficult. The new tests of broccoli in the framework of the 2005 food monitoring were intended to show in how far the situation of contamination had changed. To this end, 71 broccoli samples were analysed again for residues of plant protection products, elements and nitrate. 70% of samples stemmed from German production, 18% from Spain, and 8.5% from Italy.

Plant protection products

Broccoli was analysed for a total of 131 active substances and their metabolites. DTC and bromide were not considered this time, because both can stem from natural environmental sources. So, if low levels of these substances are found, these cannot or only with difficulty be correctly evaluated. Bromide is ubiquitous in soil, so that low-level findings must not necessarily result from the use of brome-containing fumigants. DTC are usually determined indirectly by measuring separated carbon disulphide. But broccoli has inherent sulphurous substances which might be converted into carbon disulphide in the course of residue analysis. Thus it cannot be decided whether carbon disulphide, in particular if found at low levels, actually originates in the application of DTC as a fungicide, or whether and to what degree it is of biogenic origin. Without these two substances, a large proportion of 75% of all samples was again free of measurable residues (see Figure 5-16). This situation is comparable to 1997, although the spectrum of substances looked for was much larger now. With 2.9%, the share of samples carrying residues above the respective maximum residue levels was clearly smaller than in 1997.

Maximum residue levels were only exceeded by residues of dimethoate and chlorpyrifos. As a whole, residue findings were limited to only 16 active substances of plant protection products. No substance was found in more than 10% of the samples. Average concentrations were very low and always under 0.01 mg/kg. The share of samples carrying multiple residues was also relatively low with 7.4%. The maximum was three substances in one sample.

Elements

Like artichokes, broccoli was also tested for contents of arsenic, lead, cadmium, copper, selenium, thallium and zinc. Lead and selenium were only quantifiable in 6% of samples, arsenic was found in 12%, and cadmium and thallium in about a third of the samples. Copper and zinc were found in nearly all samples.

Ninety per cent of concentrations were below 0.6 mg/kg with copper, below 6.8 mg/kg with zinc, and always below 0.03 mg/kg with the other elements. Lead and cadmium contents have halved compared to 1997 and are now on a very low level. Legal maximum levels were not exceeded.

Nitrate

Compared to 1997, when the average nitrate levels in broccoli were found in the range between 500 and 1000 mg/kg, levels found in 2005 were reduced by nearly half. The average level was 236 mg/kg, and the 90th percentile relatively low with 490 mg/kg.

Conclusion

Broccoli was again shown to be a vegetable with only slight contamination with heavy metals, nitrate, and plant protection products.

5.10
Fruiting vegetables

French beans

Beans are rich in vitamins, protein, and minerals. They are a valuable food in all cultures and not least therefore, they are on the list of foods to be regularly checked in the framework of the co-ordinated Community monitoring programme. In the framework of the national food monitoring, beans were examined before in 1995, 1996, and in 2002. Study results from the year 2002 showed medium-degree contamination with plant protection product residues and nitrate, while levels of heavy metals were very low.

In 2005, 131 samples of fresh and deep-frozen French beans were analysed for plant protection product residues, elements, and nitrate. About half of the samples stemmed from Germany, the rest from other countries, including the Netherlands, Poland, and Egypt.

Plant protection products

The spectrum of examination extended over 131 active substances and metabolites of plant protection products. Even though the range of substances looked for had been extended and the sensitivity of analytical methods improved since 2002, the proportions of samples with and without measurable residues, and with residues above maximum residue levels in 2005 were the same as in 2002. This is illustrated by Figure 5-18. This means there has been no change about the medium degree of contamination. Overall, residues of only 27 active substances were found. The most frequent finding was vinclozolin in 23.7% of samples, the same as in 2002. Only the average concentrations of vinclozolin and azoxystrobin slightly exceeded 0.01 mg/kg. The MRL non-compliance rate was 6.9%, and consisted in four cases with chlorothalonil (samples of Polish origin), two cases with fludioxonil, and one case each involving azoxystrobin, chlorpyrifos, dithiocarbamates, dimethoate, endosulfan, fenhexamid, and tolylfluanid. The proportion of samples with multiple residues was relatively small with 14.5%. The maximum number of five different residues was found only in one sample.

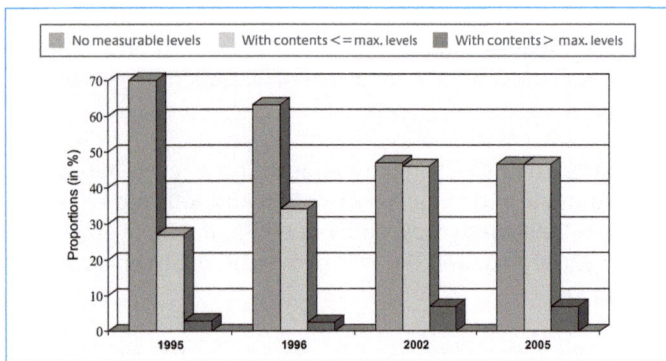

Figure 5-18 Residues of plant protection products in French beans compared by years.

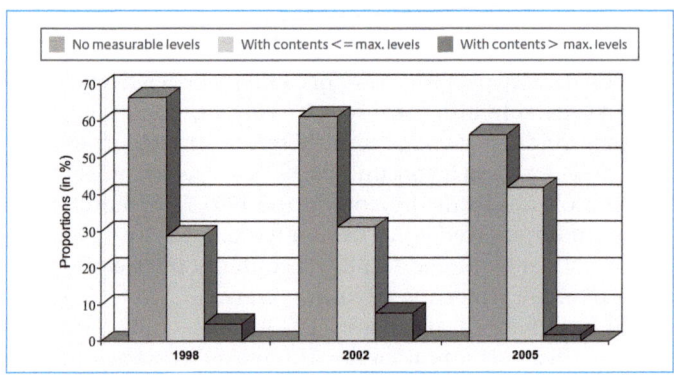

Figure 5-19 Residues of plant protection products in carrots compared by years.

Elements

French beans were analysed for contents of arsenic, lead, cadmium, copper, selenium, thallium, and zinc. Thallium was found only in one sample (0.8%) and arsenic in three (2.3%). Also relatively rare were finding of lead (10.8% of samples) and selenium (9.2%). Cadmium was quantified in nearly a quarter of the samples, and copper and zinc in nearly all. Actual element concentrations, in particular of the heavy metals, were very low, as in the previous studies of the years 1995, 1996, and 2002, and were roughly similar to the findings of the year 2002. 90% of the concentrations of arsenic, cadmium, selenium, and thallium were below 0.02 mg/kg. The 90th percentile concentration was around 0.04 mg/kg in lead, 1.0 mg/kg in copper, and 3.7 mg/kg in zinc. Legal maximum levels were not exceeded.

Nitrate

The average nitrate content was 363 mg/kg, and the 90th percentile 678 mg/kg, that is medium-degree contamination. The findings were nearly the same as in 2002.

Conclusion

Nothing has changed about the medium degree of contamination of French beans with resides of plant protection products and nitrate since the 2002 monitoring studies. Yet, only two active substances had average residue concentrations of slightly above 0.01 mg/kg. Contamination with heavy metals is overall low.

5.11
Root vegetables

Carrots

Carrots are rich in dietary fibres, minerals and fat-soluble β-carotin, a pre-stage of vitamin A. Carrots are grown as more than 60 breeds and hundreds of varieties worldwide, with an annual harvest of about 13 million tonnes.

Being a popular and much-consumed vegetable, carrots were previously subject to monitoring studies, namely in the years 1998 and 2002, with regard to pesticide residues, elements and nitrate. At that time, it was found that contamination of carrots with unwanted substances was pleasingly low.

In 2005, another 105 carrot samples were tested for the same substance groups upon the Commission's recommendation for the co-ordinated Community monitoring programme. More than three quarters of samples were of domestic origin, another 10% from Italy, and near to 7% from the Netherlands.

Plant protection products

The share of samples without quantifiable residues has slightly decreased compared to 1998 and 2002 (see Figure 5-19). This might be attributable to improved analytic techniques and an extended and more appropriate spectrum of substances analysed. However, maximum residue levels were only exceeded to very small degree in 1.9% of the samples. This concerned diniconazole and ethoprophos. Of the 131 substances analysed, only 20 were detected, including frequent findings of two fungicides, difenoconazole (24%) and azoxystrobin (12%). Average residues were only in rare cases slightly higher than 0.01 mg/kg.

Multiple residues occurred only in 16 samples, that is 15.2%. Only one sample carried the maximum number of four residues.

Elements

Carrots were analysed for arsenic, lead, cadmium, copper, selenium, thallium and zinc. Arsenic and thallium were quantified in only 13% and 7% of the samples, respectively, selenium in a quarter and lead in half of all samples. Cadmium, copper and zinc were detected in nearly all samples. As in previous years, actual concentrations were generally low. 90th percentile concentrations of arsenic, lead, cadmium, selenium, and thallium were below 0.04 mg/kg, and there was no case of non-compliance with maximum levels.

Nitrate

Nitrate levels in carrots were low also in 2005, and were comparable to those measured in 2002. The average nitrate content has been declining since 1998 and amounted to 135 mg/kg in 2005. The 90th percentile was 323 mg/kg.

Conclusion

Carrots are only lightly contaminated with residues of plant protection products, heavy metals and nitrate. Non-compliance cases with maximum residue levels of plant protection

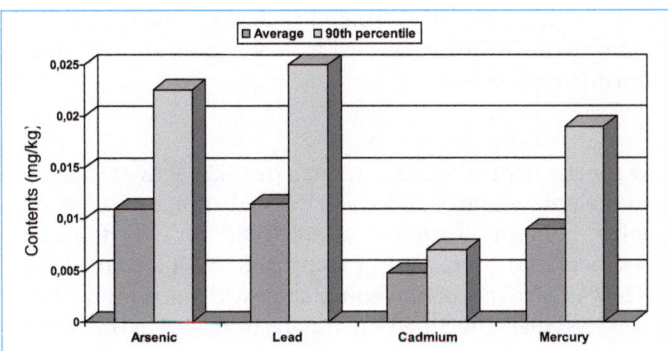

Figure 5-20 Element contents in tinned champignon.

products have clearly declined compared to 2002 and did not occur again with heavy metals. Nitrate loads have continually declined since 1998.

5.12
Mushroom products

Tinned champignon, dried shiitake

Cultivated champignon is economically the most important mushroom for human consumption and is offered throughout the year. The Far-East Shiitake mushroom is also offered throughout the year, and is becoming increasingly popular. Shiitake has already been cultured as mushroom for human consumption and for curative purposes for some 1000 years in China and Japan. Normally, cultured mushrooms such as champignon and shiitake are only exposed to controlled environmental conditions, and should therefore not carry any conspicuous levels of heavy metals or radioactive substances like wild mushrooms, provided that no contaminated substrates are used in mushroom cultivation.

Because of their economic importance, cultivated champignon was studied before in the 1999 monitoring scheme. At that time it was analysed for plant protection product residues, heavy metals and nitrate. As it had been expected for cultured

champignons, contamination levels were low overall. In the 2005 monitoring, examination of tinned champignon and dried shiitake was therefore focussed on element levels. The 82 samples of tinned champignon mainly came from Germany (45%), the Netherlands (24%), and France (13%), while the 75 dried shiitake samples stemmed from Germany (≈50%), China (24%), and Vietnam (12%). The remaining samples were of different other origins.

Elements

The mushrooms were analysed for the elements arsenic, lead, cadmium, copper, mercury, selenium and zinc. Tinned champignon was additionally tested for tin which could have been migrated to the product from the tin.

Table 5-3 compares frequency of findings and average levels of the elements found in the two mushroom products. In contrast to the findings in tinned champignon, the elements analysed were very frequently or always found in dried shiitake, and at noticeably higher concentrations. That is partly attributable to the fact that drying leads to an increase in concentration. In contrast to that, one has to take account of a certain dilution factor when considering the tinned champignon, because the analysis did not only include the actual champignon (drained weight) but the total content of the tin as a ready-to-eat food.

Figure 5-20 shows average and 90th percentile contents of arsenic, lead, cadmium and mercury in tinned champignon.

Average element contents in tinned champignon were generally low and comparable to those found in fresh cultivated champignon in 1999. Yet, 10% of the samples contained mercury at between 0.02 and 0.05 mg/kg, which is a relatively high level in vegetable food.

Apart from that, tinned champignon was found to be medium-degree contaminated with tin. Tin concentrations varied very strongly, the average concentration being 31.1 mg/kg, while the median value was 0.86 mg/kg and the 90th percentile concentration 155.4 mg/kg. The maximum concentration was 251 mg/kg and exceeded the fixed maximum level of 200 mg/kg for tin in tinned foodstuffs. That maximum level was exceeded in three samples (3.8%). The number of samples was too low to derive any correlation between the origin of sample and non-compliance with the maximum level.

	Samples with quantifiable element contents (%)		Average content (mg/kg)	
	Champignon, tinned	Shiitake, dried	Champignon, tinned	Shiitake, dried
Arsenic	18.7	96.0	0.011	0.332
Lead	14.6	93.3	0.011	0.160
Cadmium	63.4	100	0.005	0.823
Copper	100	100	1.277	8.571
Mercury	55.3	83.6	0.009	0.026
Selenium	78.7	87.8	0.054	0.145
Zinc	100	100	3.721	70.249
Tin	62.8	–	31.140	–

Table 5-3 Shares of mushroom product samples with quantifiable element contents.

When interpreting element findings in shiitake (see Figure 5-21), one has to bear in mind that dried mushroom is not consumed directly but only after soaking with a meal. If one takes account of that "dilution factor" of roughly 1/12, average element levels in shiitake can be generally rated as low. Yet, eight samples (11%) carried cadmium concentrations which would still top the maximum level of 0.2 mg/kg established for cadmium in fresh cultivated mushrooms after giving account to the above mentioned dilution factor.

Conclusion

Referred to average levels of heavy metals in fresh mushrooms, the contamination of shiitake and champignons is generally low. Yet, tinned champignon was contaminated with tin to medium degree, and dried shiitake had some conspicuous findings of increased cadmium levels. Contamination of mushroom products with heavy metals should therefore continue to be monitored, the same as contamination of fresh mushrooms intended for processing. It must also be made a general requirement that substrates used for mushroom cultivation are free from, or at least poor in, heavy metals, and that contamination during processing, for instance during the tinning process, is kept to a minimum.

5.13
Pome fruit

Pear

Pears are very popular in Germany. They are sweeter than apple and easily digestible and wholesome, among other things because of their low acid content. This makes them also ideal as dietary food. Imports, including from the southern hemisphere, allow a market offer of pears throughout the year.

Pears were examined in the framework of monitoring studies before, namely in the years 1998 and 2002, for the presence of residues of plant protection products and elements. The past studies have shown that contamination with plant protection products is of medium degree, while contamination with heavy metals is very low.

In 2005, another 108 samples of pear were analysed for the same substance groups as above according to the recommendations for the co-ordinated Community monitoring pro-

gramme. One quarter of samples each stemmed from Germany and Spain, 18% from Italy, 10% from South Africa, and the rest from other countries.

Plant protection products

The spectrum of analyses extended over 133 active substances and their metabolites. In spite of this wide range of substances analysed, the proportion of samples without detectable residues increased again slightly compared to 2002, namely from 3.8 to 7.4%. The rate of non-compliance with maximum residue levels has clearly declined since 2002, from 7.8% at that time to 4.6% now.

Residue findings in pears were related to 58 active substances. The following substances were found frequently: dithiocarbamates (in 54% of the samples), chlorpyrifos (28%), captan (24%), tolylfluanid (23%), chlormequat (19%), azinphos-methyl (16%), and chlorpyrifos-methyl and procymidone in 10% of the samples each. Non-compliance with maximum residue levels occurred with fenhexamid (twice), and with acetamiprid, lufenuron, and metoxyfenozide, each with one case.

Residue concentrations were generally low, and only in rare cases higher than 0.01 mg/kg.

A large portion of samples of 78% carried multiple residues, one sample even carried eleven and one twelve residues at a time.

Elements

Pears were analysed for contents in arsenic, lead, cadmium, copper, selenium, and zinc.

Selenium was quantifiable in 5 samples only (4.6%), and arsenic in 13 samples (12%). Lead was quantified in around a quarter and cadmium a circa half of the samples, while copper and zinc were present in nearly all samples.

Similar to the findings in 1998 and 2002, contamination of pears with heavy metals was found generally low in 2005.

90% of the findings of arsenic, lead, cadmium, and selenium were near to or slightly above 0.02 mg/kg. Even the maximum concentrations of copper and zinc were relatively low with 2.3 and 3.5 mg/kg, respectively. One single sample exceeded the maximum level of lead of 0.1 mg/kg slightly.

Conclusion

Contamination of pears with residues of plant protection products and heavy metals was only low. Cases of non-compliance with legal maximum levels have clearly declined since 2002.

5.14
Stone fruit

Peach, nectarine

Peach and nectarine are closely related botanically. Both kinds of fruit are very popular because they are very juicy, aromatic, and abundant in vitamins and mineral substances. At the same time, the proportion of fibre substances is relatively low. Peaches and nectarines are imported nearly throughout the year, with a peak from June to September.

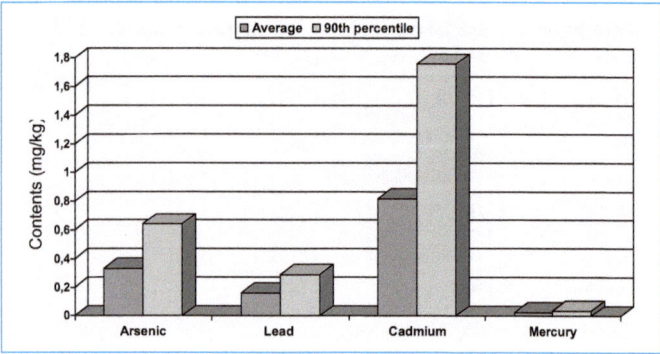

Figure 5-21 Element contents in dried shiitake.

Peaches and nectarines were tested for pesticide residues and elements before in the framework of the 1998 and 2002 monitoring studies. In 2002, it was found that contamination with heavy metals was encouragingly low. Non-compliance with maximum residue levels of plant protection products was rated as low in nectarines and of medium-degree in peaches.

In 2005, another 98 samples of peaches as well as 41 nectarine samples were examined according to the Commission's recommendations to the co-ordinated Community monitoring programme. 42% of the samples each stemmed from Italy and Spain, near to 9% from France, and the rest from other states. This allows drawing a comparison both between residue situations in the different years and between Italy and Spain as countries of origin.

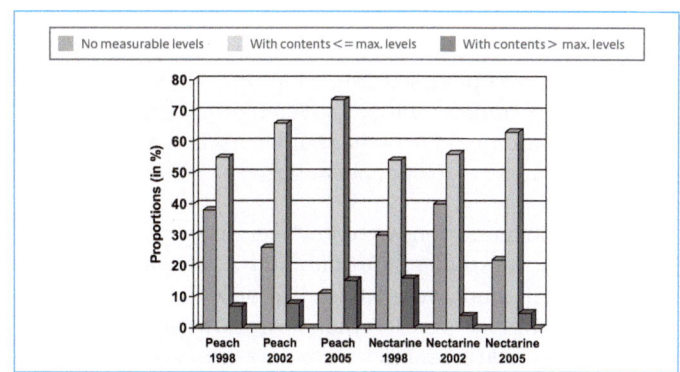

Figure 5-22 Plant protection product residues in peaches and nectarines in a comparison of years.

Plant protection products

Figure 5-22 shows that the proportion of samples with residues has considerably increased over 1998 and 2002, both in peaches and nectarines. This may be partly attributable to improved sensitivity of detection methods, and to the extended spectrum of substances searched for. But the proportion of non-compliant samples has also increased, in nectarines slightly, but considerably in peaches, with an increase from 8% in 2002 to 15.3% in 2005. This rate of non-compliance with maximum residue levels is nearly double the rate of 2002.

Residues of a total of 68 active substances of plant protection products were found in peaches and nectarines, out of a 131 active substances looked for. Frequently quantified substances were fungicidal dithiocarbamates (present in 32% of the samples), carbendazim (23%), etofenprox and iprodione (16% each), chlorpyrifos (15%) and tebuconazole and captan in 12 and 11%, respectively. Table 5-4 lists frequencies of detection of these substances in a comparison between samples originating from Italy and Spain.

This table allows certain conclusions about the frequency of application of various plant protection products in Italy and Spain.

Residue levels were generally low. Apart from dithiocarbamates and iprodione with 0.04-0.05 mg/kg, concentrations were on average mostly below or sometimes around 0.01 mg/kg.

Substances most frequently involved in non-compliance with MRLs were acrinathrin (six times) and acephate and vinclozolin (twice each), as is shown in Table 5-5. In nectarines, fenazaquin and fenpropathrin were the only substances to exceed maximum residue levels, either substance in a single case.

More than half the samples carried multiple residues (59.7%). Most frequent were two residues, the maximum finding was ten residues in one sample once.

Substance	Proportions of peach and nectarine samples with quanitifiable residues (%) listed by origin	
	Italy	Spain
Dithiocarbamates	17.5	43.1
Carbendazim	8.6	39.3
Etofenprox	35.0	0
Iprodione	5.2	20.7
Chlorpyrifos	19.0	10.3
Tebuconazole	13.8	5.2
Captan/Folpet	5.2	19.0

Table 5-4 Frequency of detection of plant protection product residues in peaches and nectarines compared by origin.

Origin	No. of non-compliance cases	Active substances concerned	
		in peach	in nectarine
France	4	Acrinathrin (3x), vinclozolin	
Italy	6	Acephat, acrinathrin, bupirimat, diazinon, vinclozolin	Fenazaquin
Spain	6	Acephat, acetamiprid, acrinathrin, diniconazole, thiabendazole	Fenpropathrin
Egypt	2	Ethoprophos, thiamethoxam	
Unknown	1	Acrinathrin	

Table 5-5 Non-compliance with MRLs, and origin of samples.

Elements

Peaches and nectarines were analysed for contents in arsenic, lead, cadmium, copper, selenium, and zinc. Arsenic was quantified in three samples only (2.3%). Selenium was found in 10%, lead and cadmium in 14% of the samples each, that is, findings were also relatively rare. Copper and zinc were found in nearly all samples.

Actual concentrations were low overall, as in the 1998 and 2002 monitoring studies. 90% of findings of arsenic, lead, cadmium and selenium were 0.02 mg/kg or less. In Copper and zinc, 90% of findings were less than 1.6 or 2.0 mg/kg, respectively. Only one nectarine sample from Italy slightly exceeded the maximum level of lead of 0.1 mg/kg.

Conclusion

Peaches and nectarines are hardly contaminated with heavy metals. The proportion of samples with residues of plant protection products, on the other hand, has considerably increased compared to earlier monitoring studies. Average residue concentrations, however, are very low. With regard to the rate of non-compliance with maximum residue levels, nectarines had only few findings, while the situation in peaches was rather bad. The residue situation in peaches should rather be improved by suitable minimisation measures.

5.15
Citrus fruit

Orange

Oranges count among the most consumed fresh fruit and is the world's most grown citrus fruit. It is popular for its taste and balanced content of various vitamins, including much vitamin C, and mineral substances.

Oranges were therefore analysed before for residues of plant protection products and elements in the 1996, 1998, and the 2002 monitoring studies. Residues of plant protection products are usually analysed according to the provisions of the German Regulations on Maximum Residue Levels, that is, for the whole fruit including peel. The intention of that prescription is to be able to check whether legal maximum residue levels have been met. Comparative studies with peeled and unpeeled oranges in 2002 confirmed the assumption that the fruit flesh, which is the part of the fruit which is actually consumed, is very little contaminated with plant protection product residues, because these stay on the peel. Contamination with heavy metals has always been very low.

In the framework of the co-ordinated Community monitoring programme, another 119 samples of oranges were analysed for pesticide residues in the whole fruit including peel and for traces of elements in the fruit flesh in 2005. Nearly half of the samples originated from Spain, 27% from South Africa, 6% from Italy, and the rest from other states.

Plant protection products

In 2005, oranges were analysed for 133 active substances and their metabolites. Residues of 55 active substances were quantifiable. Table 5-6 lists the frequently found substances and makes a comparison to findings in tangerines (see following section).

As in the previous studies, the substances imazalil, orthophenylphenol and thiabendazole, which are used for preservative surface treatment after harvest, were found very frequently, and in the case of imazalil in nearly every sample. These substances were also found with the highest average concentrations. These were 0.87 mg/kg, 0.13 mg/kg, and 0.21 mg/kg, respectively. All other residues were minor, remaining under 0.01 mg/kg, apart from very few exceptions.

Maximum residue levels were exceeded in 14 samples (11.8%). Pyriproxyfen in oranges from South Africa accounted for eight of these cases alone (6.7%). The maximum residue level here would be 0.01 mg/kg. Other cases of non-compliance with maximum residue levels were attributable to lufenuron, parathion, procymidone, prothiofos, tebufenpyrad, and thiabendazole, each in one sample.

In total, one has to state that the proportion of samples with residues above MRLs has at least doubled since 2002. Each sample carried at least one quantifiable residue, and multiple residues were present in 89% of samples. Two samples carried the maximum of eight residues.

Elements

Orange fruit flesh was analysed for the elements arsenic, lead, cadmium, copper, selenium, and zinc. Arsenic was not found at all, and selenium in three samples only. Lead and cadmium were also detected in only 21% and 12% of the samples, respectively, while copper and zinc were quantified in three quarters of the samples.

As in 2002, actual levels were generally low. 90% of the concentrations of arsenic, lead, cadmium, and selenium were below or near 0.01 mg/kg, while those of copper and zinc were around 0.5 mg/kg. Maximum levels were not exceeded.

Table 5-6 Proportions of samples with quantifiable residues in oranges and tangerines.

Active substance	Samples with quantifiable residues (%)	
	Orange	Tangerine
Imazalil	90.8	94.7
Chlorpyrifos	46.2	75.0
Orthophenylphenol	33.6	42.1
Thiabendazole	32.8	50.0
Carbendazim	21.1	30.0
Dicofol	9.5	35.0
Hexythiazox	3.7	29.4
Malathion	15.1	20.0
Dithiocarbamates	14.3	20.0
Methidathion	13.4	10.0

Tangerine

Tangerines are the second most important citrus fruit on the world market, following oranges. Their taste is less sour than that of oranges, at the same time they are as healthy and easier to peel.

Tangerines (including clementines) were studied before in the 1998 and 2002 food monitoring schemes. Similar to oranges, contamination with heavy metals was found low, while contamination with plant protection product residues was found relatively high.

The number of 20 tangerine samples taken in 2005 in the framework of the co-ordinated Community monitoring programme was relatively small. All samples came from Spain. They were analysed for plant protection product residues as whole fruit including peel according to the Regulations on Maximum Residue Levels (RHmV), but for elements only as the fruit flesh portion.

Plant protection products

The range of substances looked for included 133 active substances and metabolites. Residues of 24 substances were actually found. The substances listed in Table 5-6 were quantified frequently. As it had been expected, the substances imazalil, orthophenylphenol and thiabendazole, which are applied to the fruit surface for preservation after harvest, were most frequently detected. There were only two samples where imazalil was not present. The average concentrations of the three substances were also the highest among the residues found, with 0.62 mg/kg, 0.06 mg/kg and 0.44 mg/kg, respectively. Average concentrations of the other residues were always below 0.04 mg/kg, with the exception of chlorpyrifos and dicofol, which had average concentrations around 0.14 mg/kg.

Among the 20 tangerine samples from Spain, there was only one which slightly exceeded an MRL, namely that of flufenoxuron. That would be the same percentage as in the 1998 monitoring (5.3%), and much less than in 2002 with 16.4% of MRL non-compliance. Yet, the small number of samples in 2005 does not actually allow trend statements.

There was only one sample (5%) without residues in 2005, which would be roughly the same portion than in 2002 (3.6%).

Eighteen of the 20 samples carried more than one residue. Most frequent findings were three to five substances in one sample, the maximum of ten substances was found in one sample.

Elements

As in oranges, the fruit flesh of tangerines was analysed for arsenic, lead, cadmium, copper, selenium and zinc. Arsenic, cadmium and selenium were not found in the 20 samples. Lead was found only in four samples, while copper and zinc were detected in all.

Concentrations were again low, as in 2002, and similar to oranges. 90th percentile concentrations were 0.01 mg/kg in lead, 0.92 mg/kg in copper, and 0.7 mg/kg in zinc. Maximum levels were not exceeded.

Conclusion

Unpeeled oranges and tangerines nearly always carry residues of plant protection products, in particular such used for preservative surface treatment after harvest. The rate of non-compliance with MRLs is increased in oranges and of medium degree in tangerines. However, previous studies in the framework of the 2002 national food monitoring showed that the fruit flesh as the edible portion of the fruit is hardly contaminated, the majority of residues staying on the peel. Contamination of the fruit flesh with heavy metals is also very low.

5.16
Fruit juices

Apple juice, pineapple juice, grapefruit juice

Fruit juices 100% consist in the juice and flesh of the respective fruit. They are free of any additives such as dyes or preservatives and are therefore similarly wholesome as the fruit from which they are made. The most popular fruit juices in Germany are apple juice, followed by orange juice. Both were repeatedly analysed in the framework of monitoring studies before, and it was always found that, in contrast to the whole fruits, the juices are very little contaminated with plant protection product residues, and that heavy metal levels in orange juice are also low. The monitoring studies of 1995 and 1996, however, produced findings of the mycotoxin patulin, partly above the now valid maximum level of 0.05 mg/kg, in apple juice, obviously depending on the fungal infestation of processed apples (see Table 5-7).

A new analysis of 119 samples of apple juice in 2005 was intended to show the current situation of contamination with patulin and with elements. At the same time, the monitoring studies for the first time included pineapple juice (51 samples) and

Table 5-7 Patulin contamination of apple juice in a comparison of years (ML: Maximum level).

Year	Number of samples	Samples with patulin		Patulin level (mg/kg)		Samples exceeding ML (0.05 mg/kg)
		Number	%	90th percentile	Maximum	
1995	289	16	5.5	*	0.074	1
1996	207	31	15.0	0.005	0.067	1
2005	110	24	21.8	0.017	0.080	1

* There were no findings of patulin in more than 90 % of samples.

grapefruit juice (65 samples), which were analysed for element contents. More than 80% of the juice sampled was produced in Germany, while the actual origin of fruit and concentrate used for production was not known.

Patulin

Patulin was only analysed in apple juice samples, because contamination is probable here. Table 5-7 shows frequency and degree of contamination with patulin in 2005 and in earlier studies.

Although there was only one case of non-compliance with the maximum level again, actual contamination with patulin with regard to frequency of findings and concentration had worsened compared to previous studies. However, this does not really allow trend statements over the period from 1995 to 2005, for two reasons. First, fungal infestation of apples is different every year, depending on the weather, and therefore also the degree of contamination with that mycotoxin. Second, improved analytic techniques with finer sensitivity are now able to quantify much smaller amounts of this substance.

Elements

All juices were analysed for levels of arsenic, lead, cadmium, copper, selenium, and zinc. Table 5-8 compares proportions of samples with quantifiable amounts of elements in the three kinds of juice.

Frequency of findings of the various elements was rather similar in apple juice and pineapple juice, and in some elements quite different from the findings in grapefruit juice, in particular for arsenic and cadmium. In grapefruit juice, share of samples with quantifiable element levels are comparable with those in oranges.

Actual concentrations were low throughout. 90% of concentrations of arsenic, lead, cadmium, and selenium were below or around 0.02 mg/kg, and those of copper and zinc in the ranges of 0.3–0.6 mg/kg and 0.5-1.1 mg/kg. Only one sample of apple juice slightly exceeded the maximum level for lead of 0.05 mg/kg.

Conclusion

Juices of apple, pineapple, and grapefruit are only slightly contaminated with heavy metals. Apple juice in 2005 often contained patulin, like before in 1996. Though concentrations were generally low, they were somewhat higher than in previous studies. One sample was found to exceed the maximum level. In apple juice production it is of particular importance that no spoilt fruit are processed.

5.17

Wines

Partially fermented grape must, quality sparkling wine

Grape must is the initial product in wine production and not comparable with grape juice, which has been made stable against fermentation. Hermetically sealed, must starts to ferment relatively quickly under the influence of yeasts naturally occurring on the grape peel. As long as the must is in the process of fermenting, the partially fermented products are called *Federweißer* (white grapes), *Roter Rauscher* (red grapes), *Suser, Neuer Wein* (in Palatinate), or *Neuer Süßer* (region of South Baden).

If fermentation of grape must or secondary fermentation of wine after addition of sugar and pure culture yeast takes place in pressure-resistant tanks or bottles, carbonic acid (carbon dioxide) formed during the process cannot excape, so that a sparkling wine is produced. According to EU regulations, quality sparkling wines and *Sekt* (particular sorts of sparkling wines in Germany) are identical and must meet certain quality criteria.

Food monitoring studies in 2001, 2002, and 2004 showed that grape juice in particular, but also white and red wine, are contaminated with the mycotoxin ochratoxin A (OTA), although that substance is both degraded during fermentation and precipitated together with the yeast lees. To get a perfect overview of OTA contamination in wine and sparkling wine production, the 2005 food monitoring looked into partially fermented grape must as the initial product and quality sparkling wine as a final product. Both product groups were also checked for levels of various elements. Grape must was also analysed for patulin. The number of grape must samples totalled 75, nearly all from domestic production (93%). Quality sparkling wine samples amounted to 138, with 72% of samples from Germany and another 16% from Italy.

Ochratoxin A

The findings confirmed that OTA is degraded during fermentation. While grape juice was found to contain OTA in 70–75% of samples in the 2002 and 2004 (project 04) monitoring studies,

Table 5-8 Frequency of findings and average levels of elements in fruit juices.

Element	Samples with quantifiable levels (%)			Average levels (mg/kg)		
	Apple juice	Pineapple juice	Grapefruit juice	Apple juice	Pineapple juice	Grapefruit juice
Arsenic	10.1	16.9	2.0	0.008	0.007	0.009
Lead	7.1	4.6	11.8	0.009	0.009	0.012
Cadmium	-	-	33.3	<0.001	<0.002	0.003
Copper	28.3	53.8	51.0	0.232	0.358	0.249
Selenium	1.0	-	2.0	0.006	<0.010	0.012
Zinc	25.3	44.6	76.5	0.440	0.833	0.337

the mycotoxin was found again only in 36% of samples of partially fermented must (without grape berry peels), and ca. 12% of the quality sparkling wine samples. In comparison to that, OTA was found in only 7% of white wine samples in 2001, but in 51% of red wine samples in 2002.

This high rate of findings and the relatively high concentrations in red wine are partly attributable to the fact that the contaminated peel of the grape berries is added to the mash for colouring.

Figure 5-23 shows that OTA levels in grape must were in the range of findings in grape juice in 2002 and 2004, while levels in quality sparkling wine were comparable to those in wines.

The legal maximum level of 2 µg/kg was exceeded in one sample each with relatively high concentrations of 20 µg/kg in grape must and 10 µg/kg in quality sparkling wine.

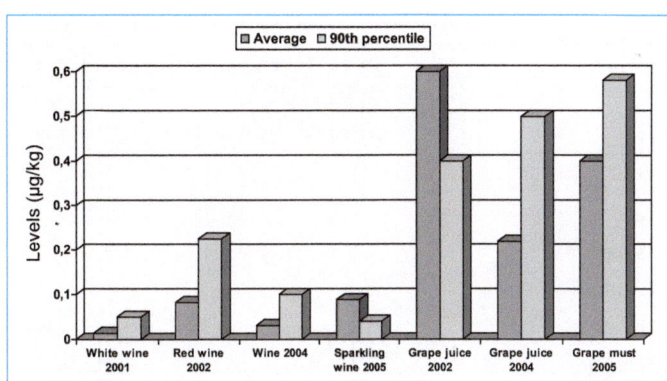

Figure 5-23 OTA levels in grape juice, partially fermented grape must, wine, and quality sparkling wine in a comparison of years. (For comparison: the maximum permissible level is 2 µg/kg).

Patulin

Patulin is a mould fungal toxin and was found in 19% of grape must samples. 90% of the concentrations found were below 12 µg/kg. The legal maximum level of 50 µg/kg was not exceeded. 21 samples of the quality sparkling wine were also tested for patulin, but none was found to contain the substance, in line with what had been expected.

Elements

Partially fermented grape must and quality sparkling wine were analysed for contents in aluminium, arsenic, lead, cadmium, copper, nickel, selenium, zinc and tin. Table 5-9 shows sample portions with quantifiable amounts of the elements in both product groups.

Findings of cadmium in particular, but also of selenium and zinc were relatively rare. Lead and zinc, aluminium in sparkling wine, and copper in grape must were frequent findings.

Quantified levels of aluminium, lead, cadmium, zinc, and tin, and of copper in sparkling wine were low. Copper had a much higher 90th percentile concentration in grape must (2.3 mg/kg) than in quality sparkling wine (0.25 mg/kg), becau-

se during fermentation it is largely reduced through adsorption to the yeast.

90th percentile concentrations of the other elements were as follows: below or around 0.01 mg/kg with arsenic, cadmium and selenium, around 0.4 mg/kg with lead, between 0.05 and 0.1 mg/kg with nickel, 0.1-0.5 mg/kg with tin, around 1.3 mg/kg with zinc, and 1.9-2.5 mg/kg with aluminium. The legal maximum level of aluminium of 8 mg/kg was slightly exceeded in one sample.

Conclusion

Contamination of partially fermented grape must and quality sparkling wine with heavy metals is only low. The mycotoxin OTA is frequently found in partially fermented grape must, similar to the findings in grape juice. In sparkling wine, OTA is relatively rare as a result of degradation during fermentation. Referred to the legal maximum level, OTA contamination is generally low. Single high-level findings above the legal maximum level (1% in both product groups) should be a reason to take more care that processed grapes are not attacked by mould. Contamination of grape must with patulin was of low degree.

Table 5-9 Frequency of findings and average levels of elements in partially fermented grape must and in quality sparkling wine.

Element	Shares of samples with quantifiable amounts (%)		Average concentrations (mg/kg)	
	Grape must, partially fermented	Quality sparkling wine	Grape must, partially fermented	Quality sparkling wine
Aluminium	36.5	62.4	0.980	1.469
Arsenic	12.7	29.4	0.005	0.008
Lead	61.9	69.8	0.020	0.020
Cadmium	1.6	0.8	0.001	0.001
Copper	77.8	25.6	0.798	0.168
Nickel	43.8	10.3	0.026	0.051
Selenium	6.3	0.8	0.005	0.005
Zinc	50.8	51.6	0.721	0.524
Tin	11.1	4.0	0.140	0.083

Table 5-10 Frequency of findings and average levels of elements in marzipan and persipan raw matters and in confectionery from other raw matters.

Element	Shares of samples with quantifiable levels (%)		Average levels (mg/kg)	
	Marzipan/ persipan raw matter	Confectionery from other raw matters	Marzipan/ persipan raw matter	Confectionery from other raw matters
Arsenic	–	2.6	<0.020	0.013
Lead	4.2	28.6	0.036	0.053
Cadmium	41.7	66.2	0.008	0.011
Copper	100	100	6.366	2.864
Nickel	83.3	90.9	0.466	0.720
Selenium	2.1	31.2	0.020	0.022
Zinc	100	100	17.240	9.152

5.18
Confectionery

Marzipan and persipan raw matter, confectionery from other raw materials

Marzipan raw matter consists in sugar, ground almonds, and sometimes rose water. Other ingredients depend on the manufacturer. Persipan is made from apricot or peach kernels instead of almonds. This gives a slightly different taste. Persipan consists of 40% ground kernel matter and 60% sugar.

Both raw matters are widely used in bakeries as components of bakery products and sweet dishes. As it is known that almonds and kernels are prone to contamination with aflatoxins, 48 samples of marzipan and persipan raw matters were analysed for contamination with these harmful mycotoxins. Nearly two thirds of the samples stemmed from domestic production.

At the same time, 77 samples of confectionery made from other raw matters were analysed for the reaction product HMF. Aflatoxins were only looked for when the product contained almonds, peanuts, or hazelnuts, for instance various cream-filled chocolate products. 80% of the sampled products are produced in Germany.

Both product groups were also examined for element contents.

Aflatoxins
Marzipan and persipan raw matter contained quantifiable levels of aflatoxin B1 in 79% of the samples, aflatoxin B2 in 25%, and aflatoxin G1 in 21% of the samples, while aflatoxin G2 was found in only 2 samples (4.2%). But even at the maximum, concentrations reached only about half the legal maximum level of 2 µg/kg of aflatoxin B1, or 4 µg/kg of total aflatoxin. The large proportion of sugar in the product causes a dilution effect, which leads to concentrations much lower than in almonds. This effect is illustrated by Figure 5-24 on the example of aflatoxin B1.

In the other confectionery, the only aflatoxin quantifiable was B1, found in only 15% of the samples. Here, too, the maximum content of 0.3 µg/kg was very low.

HMF
HMF was quantified in nearly two thirds of the confectionery samples with an average concentration of 8.6 mg/kg. The 90th percentile concentration was 24 mg/kg. Compared to that, earlier studies had produced findings between 10 and 1107 mg/kg in caramel-containing sweets such as pralines, chocolates, and candy sweets[5].

HMF was also found in 13% of samples of marzipan and persipan raw matters, but at much lower concentrations (average: 1.3 mg/kg; 90th percentile: 9 mg/kg).

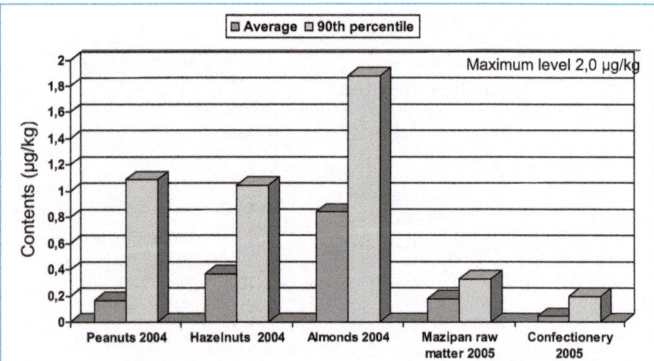

Figure 5-24 Aflatoxin B1 in shelled fruit and in sweets made therefrom.

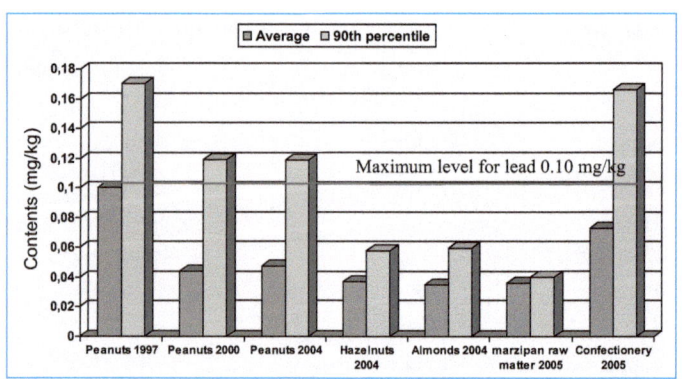

Figure 5-25 Lead contents in marzipan and persipan raw matters and in confectionery from other raw matters compared to shell fruit. (The legal maximum level is valid only for shelled fruit.)

[5] ÖKOTEST-Magazin, April 1997

Elements

Both marzipan and persipan raw matters and other confectionery were analysed for arsenic, lead, cadmium, copper, nickel, selenium, and zinc. Table 5-10 shows the proportions of samples with quantifiable levels in both product groups.

Differences in the frequencies of findings are explained by the different kinds and amounts of ingredients used in the two product groups. For instance, marzipan raw matters show a similar frequency distribution pattern as almonds. But the dilution effect through the ingredients added, in particular sugar, leads to fewer findings of elements (with the exception of copper and zinc) and generally lower concentrations, as can be roughly seen by the example of lead in Figure 5-25. 90th percentile concentrations were below or around 0.04 mg/kg with arsenic, cadmium, selenium, and with lead in marzipan/persipan raw matters, and between 0.8 and 1.5 mg/kg with nickel. 90th percentile concentrations of copper and zinc in marzipan and persipan raw matters were 11.6 mg/kg and 22.8 mg/kg, respectively, and thus much higher than in other confectionery with 4.8 and 13.9 mg/kg, respectively.

Yet, 12 confectionery samples (15.6%) displayed conspicuously high lead contents ranging between 0.12–0.49 mg/kg. This raised the 90th percentile to 0.16 mg/kg and thus relatively high (see Figure 5-25). Two cadmium levels in confectionery samples were also found higher than 0.05 mg/kg.

Conclusion

Marzipan and persipan raw matters are only to low degree contaminated with aflatoxins and heavy metals. The same holds, in principle, for confectionery of other raw matters, but these showed more often increased concentrations of lead, and in a few cases also of cadmium. The causes thereof should be identified and eliminated. HMF levels are comparatively low.

6 Results of monitoring projects

The 2005 food monitoring scheme included the ten following projects (P01 to P10) looking into particular problems:

P01:   Furan in foodstuffs
P02:   Carbendazim in fruit juices
P03:   Glycoside alkaloids in potatoes
P04:   Lead and cadmium in some supplementary foodstuffs
P05:   Residues of plant protection products in tomatoes
P06:   Persistent organo-chlorine compounds in glasshouse cucumber
P07:   Ochratoxin A, deoxynivalenol and zearalenone in cereal flours
P08:   Cadmium in cuttlefish products
P09:   Benzo(a)pyrene in smoked fish
P10:   Herbicide residues in some kinds of vegetables

These projects were managed each by an office or laboratory of the food control authorities of the federal states (*Bundesländer*). The project reports compiled in this chapter have been drawn up by the managing offices or laboratories.

The office in charge of the respective project and the other offices and laboratories participating are named at the beginning of each project report.

6.1
Project 01: Furan in foodstuffs

Office in charge: CVUA Karlsruhe
Participating offices: LAVES Niedersachsen, LUA Speyer, SUA Wiesbaden, CVUA Freiburg, LLB Brandenburg

Furan (see glossary) was first detected in coffee in 1938. Furan may be formed in foodstuffs in the course of what is called the Maillard reaction when carbohydrates are heated. Furan is also formed when ascorbic acid, amino acids or polyunsaturated fatty acids are heated. Furan levels will be particularly high when food is roasted (for instance, coffee beans) or heated in closed systems (for instance, baby food and ready-to-eat meals).

In the framework of this project, a total of 204 food samples were analysed for furan. Sample numbers and some statistical indices of measurements in the food groups are shown in Table P01-1 below.

The measurements show that furan is present in nearly all food groups analysed (see Table and Figure P01-1). While levels in bouillon and stock products where below the limit of quantification, soups had average levels of 15.8 and 43.8 μg/kg. Average levels in liquid ready-to-eat soups were three times higher and medians five times higher than in dried soup products.

Dried food products such as dried soups are finally prepared in households. Therefore, the amount of furan which is finally ingested by a consumer depends on how the food is prepared at home. Relevant for consumer protection are therefore, on the one hand, such products which carry relatively high levels of furan, such as coffee, and on the other, foodstuffs which are directly consumed. The latter include ready-to-eat meals, with an average furan content of 34.6 μg/kg, and ready-to-eat baby and infant food, with average furan contents of 19.3 and 16.5 μg/kg. Furan contamination of infant food is particularly relevant because of the low body weight of the "target consumers". Maximum furan levels measured in baby and infant food were 41 μg/kg and 65 μg/kg, respectively.

Table P01-1 Findings of furan in foodstuffs.

Kind of sample	Number of samples	Samples with residues	Average level (μg/kg)	Median (μg/kg)	90th percentile (μg/kg)	Maximum (μg/kg)
Bouillon and stock products	4	0				
Soups, dried	14	12	15.8	10.7	50.0	54
Soups, liquid	9	9	43.8	47.0		89
Baby and infant food	5	3	4.0	2.0		13
Ready-to-eat meals for babies	35	34	19.3	17.8	36.4	41
Supplementary fruit pap or vegetable pap for babies and infants	70	60	16.5	12.2	41.0	65
Ready-to-eat meals	67	63	34.6	30.0	66.2	164

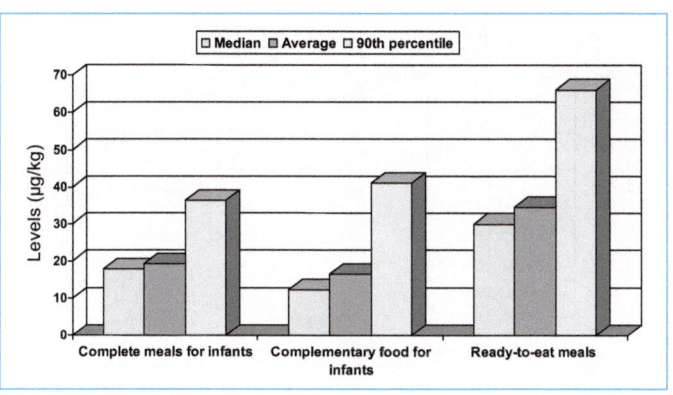

Figure P01-1 Furan levels in ready-to-eat foodstuffs.

Conclusion

The study showed that furan is present in many foodstuffs. This is important in so far as furan has been classified by the WHO as potentially carcinogenic in humans. The current state of knowledge is that the furan levels measured do not pose an acute risk to health. Yet, under the angle of preventive health protection of consumers, it is necessary to minimise furan levels, in particular in foods intended for sensitive consumer groups such as babies and infants.

6.2

Project 02: Carbendazim in fruit juices

Office in charge: CVUA Ostwestfalen-Lippe
Participating offices: LUA Sachsen, ILAT Berlin, LGL Erlangen, SUA Kassel

Carbendazim is a systemic fungicide applied to the leaves and soil to control fungal diseases in vineyards, orchards, vegetables, ornamentals, and cereals. Apart from that, it is also used after harvest in citrus fruit, stone fruit, and bananas[1]. When analysing carbendazim, it must be taken into account that this active substance at the same time is the main metabolite of the active substances benomyl and thiophanate-methyl. The German Regulations on Maximum Residue Levels therefore use the sum of the three substances (total residue expressed as carbendazim) as the measure for compliance with legal regulations. In 2005, Germany had only one plant protection product on the basis of thiophanate-methyl (Cercobin FL) authorised

for pre-harvest use against fungal storage rots in pome fruit.

Analyses of red grape juice for residues of plant protection products in the framework of the 2002 National Food Monitoring produced signs of frequent use of the fungicide carbendazim in grapes.

The present project was intended to verify the data about the situation of contamination in grape juices obtained in 2002 and, in addition to that, collect information on the contamination of pome and stone fruit juices.

Depending on the analytic method used, the detection limit for carbendazim was between 0.01 and 0.005 mg/kg with HPLC with UV or fluorescence detection, and between 0.001 and 0.0005 mg/kg with mass spectroscopy.

A total of 228 fruit juices were analysed for carbendazim. The kinds of juice, numbers of samples and proportions with findings are listed in Table P02-1 below.

While carbendazim was found in around 60% of red grape juice, the rate of findings in white grape juice was only 17% (see Table P02-1). The maximum level of carbendazim was 0.049 mg/kg in red grape juice and around 0.005 mg/kg in white grape juice. Apple juice, too, had positive findings in about 40% of the samples, with a maximum value of 0.043 mg/kg.

In pear juice, the substance was found in only one of the 25 samples analysed, while orange juice samples were free of carbendazim throughout.

The category "Others" includes two cherry juices, two currant nectars, one plum nectar, and five mixed juices with some apple. Carbendazim was found only in three samples of the mixed juices with apple.

Given a national maximum residue level (MRL) of thiophanate methyl (expressed as the main metabolite carbendazim) of 2 mg/kg, and further given the transfer factors of 1 and 0.8 for grape juice and apple juice, respectively, which have been proposed for carbendazim for the purpose of EU evaluation of active substances under Directive 91/414/EEC, it may be concluded, from the findings in juice, that the maximum residue levels were not exceeded in the fresh fruit.

It is noticeable that the highest proportion of positive findings occurred in red grape juice. It was not possible to determine the origin of the ware as the juice samples were mainly drawn at supermarkets and discounters.

Conclusion

Carbendazim was found in a relatively large share of grape

Fruit juice	Number of samples	Samples with carbendazim	Average v. (mg/kg)	90th perc. (mg/kg)	Maximum (mg/kg)
Grape juice, red	72	44 (61%)	0.013	0.029	0.049
Grape juice, white	18	3 (17%)	<0.001	0.004	0.005
Apple juice	62	25 (40%)	0.003	0.010	0.043
Pear juice	25	1 (4%)	--	--	0.002
Orange juice	41	0 (0%)	--	--	--
Others	10	3 (30%)	--	--	0.004

Table P02-1 Carbendazim in fruit juices.

---

[1] Papadopolou-Mourkidou E. (1991), J. AOAC Int. 74: 745

and apple juices. Actual concentrations, however, were low. In orange and pear juice, the substance was not detected, or only in single cases. Although total contamination with carbendazim is low, further studies will be interesting to differentiate between both origins and ways of production, as direct juice or juice made from concentrate.

6.3
Project 03: Glycoside alkaloids in potatoes

Office in charge:	LGL Erlangen
Participating offices:	CVUA OWL Bielefeld, CUA Hagen, TLLV Bad Langensalza, LSH Kiel

The poisonous glycoside alkaloids solanine and chaconine are naturally occurring in potatoes to small amounts. To date, a total glycoside alkaloid content (sum of solanine and chaconine) of up to 200 mg/kg is believed to be harmless. The Joint FAO/WHO Expert Committee on Food Additives regards a total glycoside alkaloid content in potatoes of between 20 and 100 mg/kg as normal[2].

However, solanine and chaconine contents may be increased in particular in strongly greened, germinating, damaged, or unripe potato tubers. To reduce the content of glycoside alkaloids before consumption, consumers are normally recommended to generously cut out spots of damage and germs, and to pour away the boiling water. The main objective of the present monitoring project was to collect topical data on the contamination of food potatoes with solanine and chaconine.

As contents of glycoside alkaloids are fluctuating depending on the degree of maturity, conditions of harvest and storage, potatoes were sampled in the framework of this project at the various trade levels – at the producer's premises, at wholesale and retail markets – and seasons, to take account of the seasonal product in autumn as well as of stored product sold in winter and spring and the new potatoes of early summer.

In total, 222 potato samples were analysed. The total alkaloid content (sum of solanine and chaconine) of 92% of the samples was in the normal range of up to 100 mg/kg. Only one sample exceeded 200 mg/kg with a concentration of 271 mg/kg. Average solanine and chaconine contents lay between 20-30 mg/kg (see Figure P03-1).

Early potatoes (mostly sampled in April and June) and autumn seasonal ware, which was mostly sampled in September, had a higher rate of increased total alkaloid contents than stored potatoes sampled in the first half of the year, mostly in January and February (see Figure P03-2). The average concentration was about 59 mg/kg and thus nearly twice the average concentration in stored potatoes (32 mg/kg).

The reasons for that are manifold. Apart from lack of maturity, deficient storage ability may have a negative impact on the total alkaloid content in early potatoes and autumn seasonal ware. New potatoes are imported onto the German market as early as in January and stay on the market for a longer period

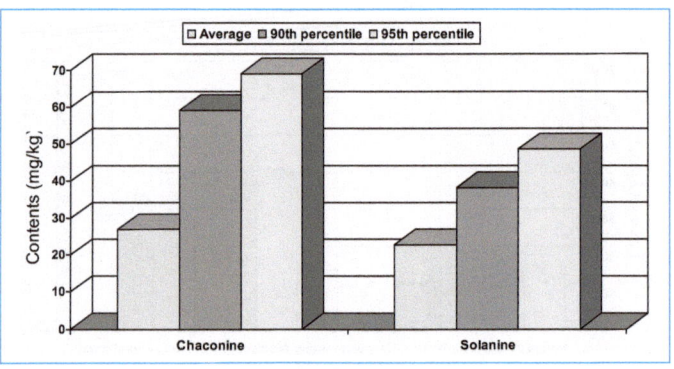

Figure P03-1 Glycoside alkaloids in potatoes.

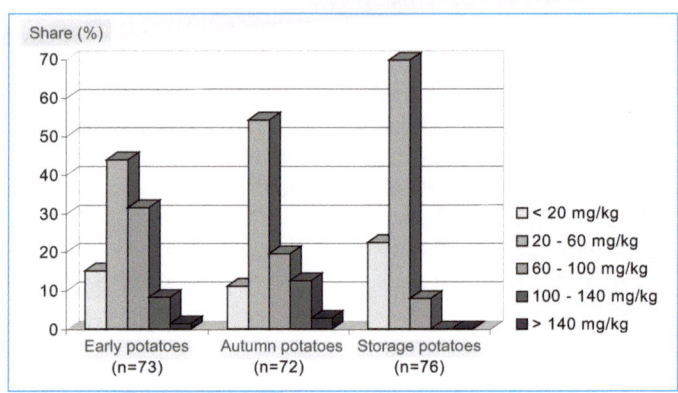

Figure P03-2 Distribution of total alkaloid contents (solanine and chaconine) as a function of season.

of time, the same as domestic early potatoes from April or May. The samples therefore did not only include fresh potatoes, but also such which had been stored for sale for some time and had developed higher levels of glycoside alkaloids. In contrast to that, potatoes which are actually intended for storage are processed with particular care, treated with germination inhibitors if necessary, and stored under optimum conditions. These factors explain the low levels of total alkaloids measured.

Potatoes show also differences in solanine and chaconine contents depending on the variety. The varieties 'Nicola', 'Cilena', and 'Princess', for instance, were sampled throughout the year and at sample numbers of more than 20. The variety 'Nicola' showed a higher average total alkaloid content than the other two, but this may be to some degree owing to the fact that 'Nicola' was mainly harvested and sampled as an early potato.

Roughly 80% of the potato samples stemmed from Germany, the rest from Italy, France, Spain, Egypt, Morocco, and others. As imported potatoes were mostly early potatoes, the foreign samples on average showed increased glycoside alkaloid contents as they are typical of early potatoes (see Table P03-1).

A comparison of findings in the different places of sampling, categorised as producers (including direct traders), wholesale traders (including packagers and importers), and retail traders (including greengrocers and food retail stores, supermarkets and market counters) shows that the share of samples with increased glycoside alkaloid contents is increasing from the producer to the retail stage (see Figure P03-3). The trend is still recognisable when the differentiation by place of

---

[2] Summary of Evaluations Performed by the Joint FAO/WHO Expert Committee on Food Additives, http://www.inchem.org/documents/jecfa/jeceval/jec_2180.htm, http://www.inchem.org/documents/jecfa/jeceval/jec_399.htm, 1992

Origin	Number of samples	Average (mg/kg)	90th perc. (mg/kg)	Maximum (mg/kg)
Germany	177	47.7	94.9	270.7
Europe and northern mediterranean region	20	59.0	79.0	122.4
North Africa and southern mediterranean region	20	61.3	109.7	137.2
Unknown	5	31.1	51.3	64.7

Table P03-1 Total alkaloid content as a function of origin.

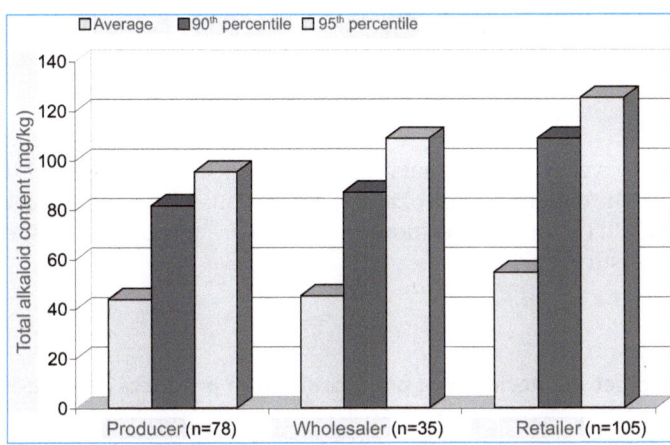

Figure P03-3 Total alkaloid content as a function of place of sampling.

sampling is made separately for early potatoes, autumn seasonal potatoes, and storage potatoes. Obviously, storage conditions can be better controlled at producers' places and wholesale stores. Yet, making allowance for the fact that most of the potatoes sampled at the retail stage were early potatoes, retail traders are not worse than producers or wholesale traders with regard to the average contents of glycoside alkaloids.

Conclusion

Glycoside alkaloid contents determined in potatoes were found to be not problematic.

The data at hand allowed to differentiate solanine and chaconine contents between early, seasonal, and storage potatoes, as well as between the places of sampling. Yet, to make well-founded statements about the differences in solanine and chaconine contents in, for instance, new potatoes and stored potatoes in the period from January to June, would require more studies with a corresponding objective of study. Such studies could be combined with analyses for residues of plant protection products, in particular germination inhibitors.

6.4
Project 04: Lead and cadmium in certain dietary supplements

Office in charge: AfV Mettmann
Participating offices: CUA Bonn, LUGV Dresden, TLLV Erfurt, LVLUA Halle, CVUA Karlsruhe, CVUA Münster, LGL Oberschleißheim, CLUA Paderborn, LVGA Saarbrücken, CVUA Stuttgart, CUA Viersen

Lead and cadmium are ubiquitous in the environment and enter foodstuffs and dietary supplements via different paths. Neither Germany nor the EU has regulated maximum levels of lead and cadmium in dietary supplements. But there have been some, though rare, notifications within the European Rapid Alerts System for Food and Feed (RASFF) about increased levels of lead and cadmium in some dietary supplements since 2002. EU experts are therefore discussing the need for an EU-wide legal maximum level for lead and/or cadmium in dietary supplements.

Analyses in the framework of this project measured lead and cadmium levels in vitamin preparations, mineral substance preparations, combined vitamin and mineral preparations, plant extract preparations, and algal preparations. A total of 306 samples were examined, both from German and foreign manufacturers, while no significant differences between domestic and foreign products were found. The analytic limit of quantification which had to be met as a minimum was 0.1 mg/kg for lead and 0.01 mg/kg for cadmium, each related to the kind of matter as which the product is offered.

It was found that 180 samples (59.2%) contained quantifiable levels of lead and 171 (55.9%) quantifiable levels of cadmium.

A comparison of the different product groups shows that vitamin preparations were the ones with the lowest lead contents. Here, the highest concentration found was 0.67 mg/kg, and the 90th percentile concentration 0.11 mg/kg. Mineral substance preparations, combined vitamin and mineral preparations, and plant extract preparations carried maximum concentrations between 2.1 and 5.1 mg/kg. So, these product groups include single cases of clearly higher contamination with lead.

The situation with cadmium in the dietary supplements examined under this project is similar. Vitamin preparations, mineral substance preparations, combined vitamin and mineral preparations, and plant extract preparations displayed cadmium levels of up to 0.62 mg/kg, with vitamin preparations again being the least contaminated. This may be attributable to less contamination of the vitamin compounds used, compared to the plant extract, algae or minerals used as raw materials for the other products.

Compared to the other product groups, algal products were clearly more contaminated with lead and cadmium. In contrast to the other products, algal preparations carried quantifiable lead in 81% of the samples and detectable cadmium in 91% of samples.

Lead in 4 samples exceeded 4 mg/kg, the highest concentration being 4.5 mg/kg. Eight samples carried cadmium levels of more than 15 mg/kg, with a maximum of 23.6 mg/kg. Such contamination must be evaluated as extremely high. All these

	Number of samples	Samples with quantifiable levels	Average con-centration	Median	90th perc. conc.	Maximum
Vitamin preparations	66	17	0.060	0.050	0.113	0.670
Mineral substance preps	66	46	0.250	0.056	0.743	5.080
Vitamin and mineral substance preparations	93	61	0.235	0.050	0.418	2.170
Plant extract preps	36	21	0.228	0.051	0.761	2.130
Algal preparations, total	43	35	1.155	0.630	3.728	4.500
Total	304	180	0.329	0.050	1.083	5.080

Table P04-1 Lead contents in dietary supplements (as mg/kg).

samples were preparations of *Spirulina* algae, while different manufacturers and different production lots were concerned.

Manufacturers' consumption recommendations and labels mostly suggest that the product should be consumed over a longer period of time. So, if the conspicuous contamination with cadmium persists over a series of production batches, it cannot be excluded that consumers will be considerably contaminated over a longer period of time, and that health may be impaired.

Conclusion

The findings of this project have shown that 90% of samples of dietary supplements contain quantifiable levels of lead under 1.1 mg/kg and of cadmium under 0.3 mg/kg. Yet, single samples with higher concentrations are conspicuous. Particular attention must be given to algal preparations, which are both more contaminated with lead and cadmium in general, and carry extremely high levels of cadmium in some cases. It must be assumed that increased levels of lead and cadmium are owing to contamination of the *Spirulina* alga which is used as the raw matter, because algae accumulate heavy metals from water to a particular extent. Health impairments cannot be ruled out as a result of increased exposure to cadmium after taking in certain amounts of these algal preparations over a longer period of time as recommended by the manufacturers. Given the fact that it is obviously technologically feasible to produce preparations with comparably low levels of lead and cadmium, and that, on the other hand, some preparations stand out with clearly higher contamination, it seems to make sense to try and reduce contamination to the technically feasible and unavoid-

able level by setting legal maximum levels.

The contamination of algal preparations with heavy metals should continue to be monitored in the framework of routine food surveillance.

6.5
Project 05: Residues of plant protection products in tomatoes

Office in charge: LAVES-LI Oldenburg
Participating offices: LGL Erlangen, LAV Halle,
                CVUA Münster

Food monitoring studies in 2001 and 2004 showed that tomatoes were contaminated with plant protection product residues to medium degree. The present project had the objective to examine tomatoes of various origins and types of growing for an extended range of plant protection products.

In total, 215 tomato samples were examined for 100 different plant protection products at least. Residues were found in 85% (148 samples) of the 175 samples stemming from conventional growing and in 78% (31 samples) of the 40 samples which were labelled as ecologically grown. This means that only 17% of all tomato samples did not carry any measurable residues of plant protection products (see Table P05-1). 55% (n = 96) of the conventionally grown samples and 8% (n = 3) of the samples labelled as 'ecologically grown' carried residues of more than one active substance. In total, 46% (n = 99) of all samples carried more than one residue (see Figure P05-1), and 16% (n = 34) of samples carried five or more residues. There may be different causes for these

	Number of samples	Samples with quantifiable levels	Average conc.	Median	90th Percentile	Maximum
Vitamin preparations	66	12	0.016	0.005	0.027	0.192
Mineral substance preps	67	39	0.105	0.039	0.302	0.441
Vitamin and mineral preps	94	60	0.084	0.015	0.270	0.580
Plant extract preparations	36	21	0.052	0.006	0.108	0.618
Algal preparations, total	43	39	4.185	0.144	21.736	23.600
Total	306	171	0.646	0.010	0.310	23.600

Table P04-2 Cadmium contents in dietary supplements (as mg/kg).

multiple residue findings: apart from possible non-compliance with good agricultural practice, multiple residue findings may also be attributable to specific measures to avoid development of resistance in pests, forestall exceeding of MRLs, selectively control different pests, or mutually enhance the efficiency of the active substances applied. It is also possible that various lots of pesticides are mixed and sold. Among the three countries where most of the tomato samples originated, Germany was the one with the lowest percentage of multiple residue findings, while Spain had the highest percentage of multiple findings.

The present findings show that there is a broad spectrum of active substances which continue to be used in tomato crops, as a total of 66 different active substances, mainly fungicides and insecticides, have been detected. Bromide was most frequently quantified in the samples which were analysed for that substance (see Figure P05-2). Most of the concentrations were in the range of up to 1 mg/kg, as is shown in Figure P05-3. It is, however, difficult to evaluate bromide contents as residues of brome-containing fumigants for soil and storage treatment, as the physiological concentration of bromide in tomatoes is also influenced by the soil nature and by fertilisation. So, only higher levels may be interpreted as signalling the use of brome-contain-

ing fumigants for pest control. The maximum residue level for bromide in tomatoes is 30 mg/kg. Two samples clearly exceeded that level with concentrations of 32.2 mg/kg and 39.2 mg/kg.

Fungicidal substances, such as dithiocarbamates and pyrimethanil, were quantifiable in more than 10% of the samples examined therefore. Procymidone, iprodione, chlorothalonil,

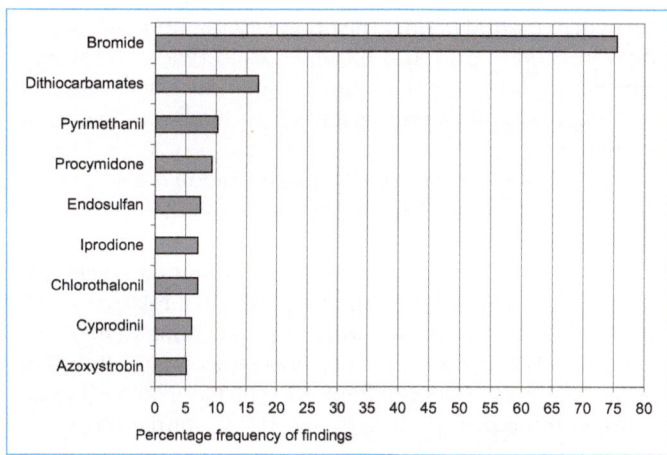

Figure P05-2 Active substances frequently quantified in tomatoes.

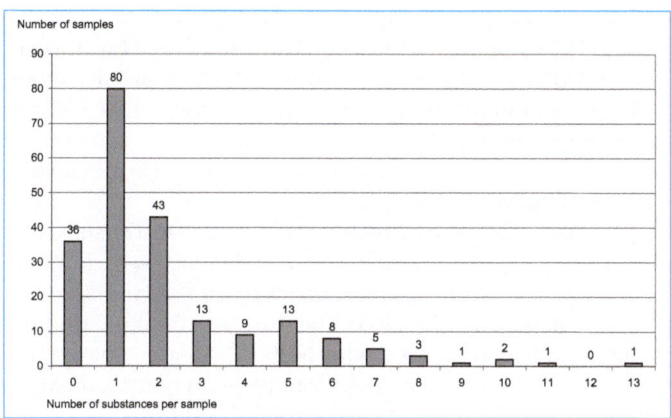

Figure P05-1 Multiple residues in tomatoes.

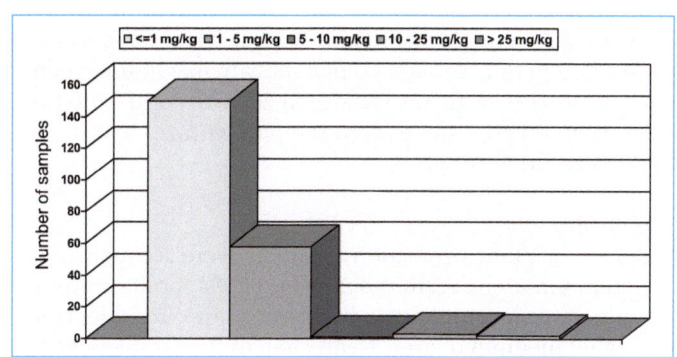

Figure P05-3 Numbers of samples containing bromide at various concentrations.

Table P05-1 Residues of plant protection products in tomatoes.

Country of origin	Number of samples	Number of samples with residues	Samples exceeding MRLs	Substances exceeding MRLs	Samples with multiple residues
Belgium	10	10 (100%)	0		5 (50%)
Germany	72	45 (63%)	4 (6%)	Acephate, bupirimate, chlormequat (twice), dimethoate/omethoate, methamidophos	13 (18%)
France	6	6 (100%)	0		3 (50%)
Israel	4	4 (100%)	0		1 (25%)
Italy	11	11 (100%)	3 (27%)	Acetamiprid, bromide (twice), chlormequat	7 (64%)
Malta	2	2 (100%)	0		1 (50%)
Morocco	4	4 (100%)	0		3 (75%)
Netherlands	56	49 (88%)	0		29 (52%)
Spain	50	48 (96%)	3 (6%)	Acetamiprid, lufenuron, orthophenyl phenol	37 (74%)
Total	215	179 (83%)	10 (5%)		99 (46%)

cyprodinil and azoxystrobin, which are also fungicides, and the insecticide endosulfan were quantified in at least 5% each of the samples examined therefore.

Residues in 10 samples (5%) exceeded MRLs in 13 cases (see Table P05-1). Most cases of MRL non-compliance (three) involved the growth regulator chlormequat. Two of the samples came from Germany, although chlormequat is authorised in Germany only for haulm stabilisation in field crops and stunting of ornamental plants. The other cases of non-compliance involved acetamiprid and bromide, each twice, the insecticides acephate, dimethoate/omethoate, lufenuron, and methamidophos, and the fungicides bupirimate and orthophenylphenol.

The samples exceeding maximum residue levels all came from Germany (four), Italy (three), and Spain (three). Tomatoes from Italy had the highest rate of MRL non-compliance, with 27%, while none of the samples from the Netherlands exceeded any MRL, although Dutch samples made up a relatively large portion. One sample from Germany even contained increased residue levels of three active substances – acephate, dimethoate/omethoate and methamidophos – which are not authorised for use in tomato crops in Germany. If this sample was consumed, a health risk to children aged between 2 and 5 years cannot be safely excluded because of the content of methamidophos at a level of 1.06 mg/kg.

If one compares the study result of 2005 with the findings of the years 2001 and 2004 (see Figure P05-4), it is striking that the proportion of samples non-compliant with maximum residue levels is nearly exactly the same as in 2004. In contrast to that, there is an obvious increase in samples with residues below MRLs over earlier studies.

Conclusion

Residues of plant protection products were frequently detected in tomatoes. As in the 2001 and 2004 monitoring programmes, non-compliance with maximum residue levels was rated medium degree. Nearly half of the samples carried more than one residue, and 16% even five or more. Germany as a country of origin had the largest share in samples without measurable residues and the smallest share in samples with multiple residues.

6.6
Project 06: Persistent organo-chlorine compounds in glasshouse cucumbers

Office in charge: LSH Neumünster
Participating offices: HU Hamburg, LAVES-LI Oldenburg, CLUA Dortmund, CEL Recklinghausen, LGL Erlangen, LUA Leipzig, TLLV Erfurt

It appears that cucumbers and other cucurbitous plants selectively accumulate persistent organo-chlorine compounds (POC) such as dieldrin and heptachlor epoxide from the soil, where these substances, owing to their extreme persistence, are still present from use as plant protection products many years ago. Temperature and in particular humidity in the soil

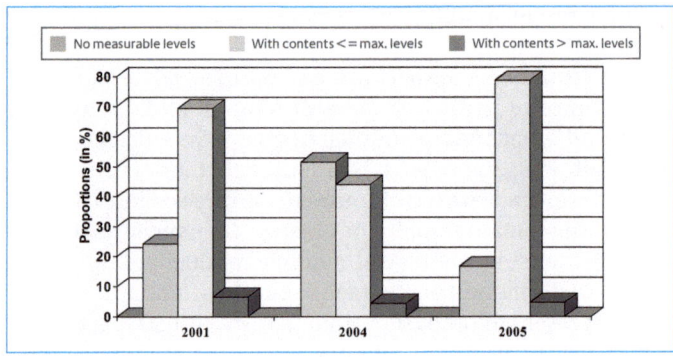

Figure P05-4 Plant protection product residues in tomatoes compared by years.

of glasshouses are different from those in the field. This may retard chemical and microbial degradation of these substances in glasshouse soils, which are often used for years. The substances are therefore present at higher concentrations than in the field. Local investigations in 2004 produced some findings above maximum residue levels in glasshouse cucumbers.

This project was therefore intended to examine domestic glasshouse cucumbers with regard to the presence of aldrin, dieldrin, heptachlor epoxide, HCH and DDT isomers, HCB and endosulfan sulphate. At the same time, the project was intended to produce data for the co-ordinated Community monitoring programme. The spectrum of substances analysed was therefore extended to current active substances in nearly all samples, while samples of foreign origin were also included in the investigations.

The project had the following findings of POC: the substances dieldrin, heptachlor epoxide, HCB and endrin were found in 45 of the 204 domestic samples analysed. These 45 positive samples included 27 samples with dieldrin, 10 samples with heptachlor epoxide, 6 samples with both substances, one sample with dieldrin and HCB, and one sample with endrin, dieldrin, and HCB. Table P06-1 gives a survey of the findings.

A comparison with maximum residue levels shows that dieldrin concentrations exceed the MRL (0.02 mg/kg) in 14% of the positive samples, and heptachlor epoxide concentrations exceed the MRL (0.01 mg/kg) in 75% of positive samples. Table P06-2 shows a breakdown of sample numbers and findings by eight regions (federal states).

On an average, 22% of samples had POC findings, 8% even above MRLs. But the regional distribution of findings is different. There are regions without POC findings, some with low and medium-degree findings, but also such with high-rate findings. Long-term use of glasshouse soils obviously differs from region to region.

In the case of the finding of endrin, dieldrin, and HCB in one sample, the soil of the glasshouse was also analysed and shown to display the same pattern of substances.

Proceeding from the above mentioned average contents of substances and an average daily consumption of 9.5 g cucumber[3], the acceptable daily intake (ADI) of dieldrin and heptachlor epoxide for a person of 60 kg body weight would be covered to 2 to 4%.

Active substance	Number of findings	Range of conc. mg/kg	Average mg/kg	Median mg/kg	90th percentile mg/kg
Dieldrin	35	0.001–0.049	0.012	0.007	0.029
Heptachlor epoxide	16	0.005–0.059	0.024	0.017	0.044
HCB	2	0.001			
Endrin	1	0.015			

Table P06-1 Organo-chlorine compounds in domestic cucumbers.

Region (federal state)	Number of samples	Number of samples with POC findings	Number of samples with POC findings above MRLs
Saxony	15	0	0
Thuringia	18	0	0
Lower Saxony	26	2 (7.7%)	1 (3.8%)
North Rhine-Westphalia	46	8 (17.4%)	1 (2.2%)
Bavaria	41	9 (21.9%)	1 (2.4%)
Schleswig-Holstein	13	5 (38.5%)	1 (7.7%)
Hamburg	26	19 (73.1%)	13 (50%)
Unknown	19	2 (10.5%)	0
Total	204	45 (22.1%)	17 (8.3%)

Table P06-2 Residue findings in cucumbers, broken down by sampling regions.

The project, which was run in the framework of the co-ordinated Community monitoring programme, examined a total of 257 samples for an average of 173 active substances (median: 204, minimum: 17, maximum: 262). 204 samples stemmed from German production, 157 of them were taken at the growers' level. Residues were found in 48% of all samples, 18% of samples carried residues of several active substances at a time, and 8% of samples exceeded MRLs. Table P06-3 shows findings for the different countries of origin.

Table P06-4 shows the distribution of multiple residues over origins.

42 different active substances were quantified. Figure P06-1 shows the ones more frequently quantified.

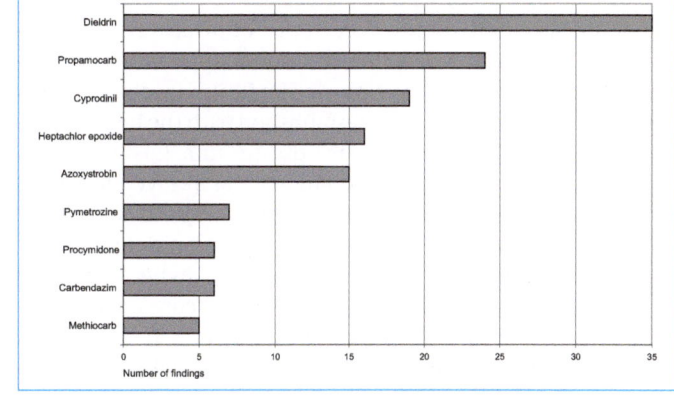

Figure P06-1 Active substances frequently quantified in glasshouse cucumbers.

Conclusion

The study has shown that residues of persistent active substances of old plant protection products may still be present in foods to considerable degree, sometimes even at levels much higher than MRLs, 25 years after these products were banned. The cause of this phenomenon must bee sought in the soil of the glasshouses, which may still be contaminated with these substances provided it is old enough and was years ago treated with the active substances in question, together with the ability of cucurbitaceans to selectively accumulate these substances. Special regional conditions also play a role. As the acceptable daily intake (ADI) of POC is by far not exhausted with average consumption, health problems are not indicated.

With regard to residues of currently used plant protection products, the situation can be evaluated as positive, as the proportions of samples with positive findings, residues above MRLs, and multiple residues of currently used products are relatively small.

6.7

Project 07: Ochratoxin A, deoxynivalenol and zearalenone in cereal flours

Office in charge: LALLF Rostock
Participating offices: LAVES-LI Braunschweig, CUA Hamm, CVUA Münster, SUAH Wiesbaden, CVUA Sigmaringen, CVUA Stuttgart, LGL Oberschleißheim, LVGA Saarbrücken

3 Schroeter A, Sommerfeld G, Klein H, Hübner D (1999) Warenkorb für das Lebensmittel-Monitoring in der Bundesrepublik Deutschland. Bundesgesundheitsblatt 1-1999, 77-83

Table P06-3 Residues in cucumbers of different countries of origins.

Country of origin	Number of samples	With residues	Samples above MRLs	Substances above MRL	Samples with multiple residues
Germany	204	99	19	Heptachlor epoxide (12x), dieldrin (5x), endrin (1x), lambda-cyhalothrin (1x), fenhexamide (1x), etridiazole (1x)	37
Netherlands	41	17	1	Etridiazole (1x), triflumizole (1x)	5
Spain	7	5	0		4
Bulgaria	2	0			
Belgium	1	0			
Unknown	2	1	0		

Table P06-4 Multiple residue findings in cucumbers.

Country of origin	Number of residues per sample							
	0	1	2	3	4	5	6	11
Germany	105	62	27	6	2	1	1	
Netherlands	23	12	3	2				
Spain	2	1	1	0	0	2	0	1
Others	3							
Unknown	1	1						

The interest in data on the occurrence of fusarium toxins, including deoxynivalenol and zearalenone, has noticeably increased over the past few years. As maximum levels for mycotoxins in cereal products were established with the Regulations on Maximum Levels of Mycotoxins, the question has arisen how the current situation of contamination looks like. While cereals intended for human consumption were repeatedly examined in the framework of the national food monitoring, cereal flours have not been examined so far. Ochratoxin A has been included in the study because it is also playing a role in cereal products.

A total of 246 samples were examined in the framework of this project. All samples were flours of rye and wheat of different degree of grinding. Samples were drawn at all levels: 51 at the production stage, 41 at the stage of manual or industrial processing, 3 at wholesale and storage places, and 151 at retail stores. The predominantly sampled rye flours were types 1150 and 997, and wheat flours types 405, 550, and wholemeal flour.

Below, findings of the mycotoxins analysed are presented in a comparison of rye and wheat flour. It did not seem practical to further break down flours by types because of the sometimes very small sample numbers.

Overall, it is to be seen that both average concentrations and 90th and 95th percentile concentrations are clearly below the respective legal maximum levels.

Ochratoxin A was the only mycotoxin where the legal maximum level of 3.0 μg/kg was exceeded, namely in three samples of type 1150 rye flour and one sample of wholemeal rye flour. This makes a share of 4% of the rye flours analysed. Quantifiable amounts of ochratoxin A were found in 44% of rye flour and 23% of wheat flour samples. The striking difference in contamination of rye and wheat flours with ochratoxin A becomes clear in Figure P07-1.

Zearalenone was not detectable in 93% of rye flour and 98% of wheat flour samples. Percentile values can therefore not be represented. Quantifiable amounts were found only in two type 997 rye flour samples and one type 405 wheat flour sample. With a maximum 14.1 μg/kg in rye flour and 10.1 μg/kg in wheat flour, the concentrations were far below the legal maximum level of 50 μg/kg.

The situation with deoxynivalenol is slightly different, if one compares cereals (see Figure P07-2). Here, rye flour is a bit less frequently contaminated than wheat flour. The share of samples with quantifiable amounts of deoxynivalenol is highest among the mycotoxins, with 27% in rye flours and 56% in wheat flours. Maximum concentrations were found in wheat

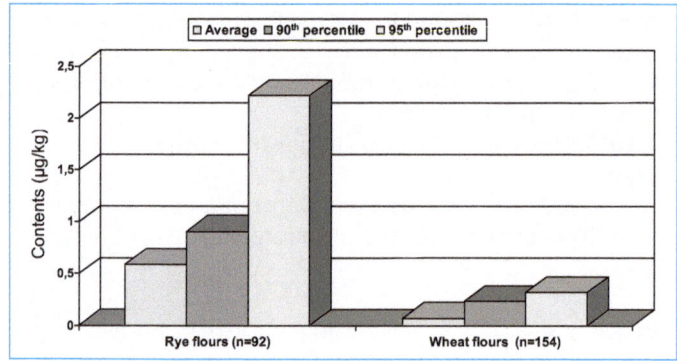

Figure P07-1 Ochratoxin A in cereal flours (legal maximum level: 3 μg/kg).

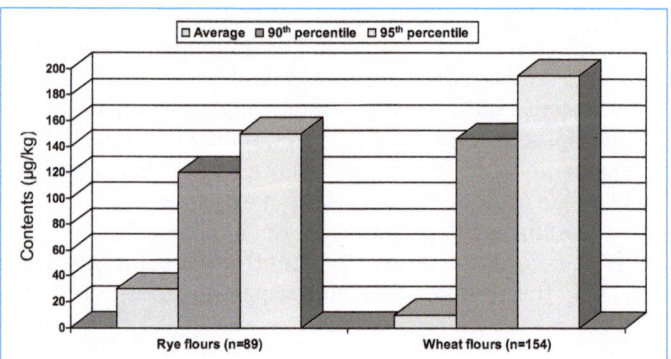

Figure P07-2 Deoxynivalenol in cereal flours (legal maximum level: 500 μg/kg).

flour. Although the legal maximum level of 500 μg/kg was not exceeded, contents measured in two wheat flour samples were hardly below that limit, with 477 μg/kg and 496 μg/kg.

Apart from the single findings, there were also 44 samples with combined mycotoxins findings. The most frequent finding was ochratoxin A together with deoxynivalenol (39 samples). Two samples carried both ochratoxin A and zearalenone, and three samples carried all three mycotoxins. There was no finding of deoxynivalenol with zearalenone alone.

As we have said, deoxynivalenol and zearalenone have been evaluated on the basis of the legal maximum levels set by the national Regulations on Maximum Levels for Mycotoxins. The EU has now made its own regulations by Regulation (EC) No. 856/2005 amending Regulation (EC) No. 466/2001. These regulations have been effective since 01 July 2006 and allow 750 μg/kg deoxynivalenol and 75 μg/kg zearalenone in cereal products. The findings of this project show, however, that the more stringent national regulations can be complied with.

Conclusion

Contamination of cereal flours with ochratoxin A, zearalenone and deoxynivalenol was low in the 2005 study period. Zearalenone does not even seem to play a role at all. Deoxynivalenol was quantified most frequently, but all concentrations were below the maximum level. The maximum level of ochratoxin A was exceeded in four samples. The findings of this project were that rye flour is more frequently and to higher degree contaminated with Ochratoxin A than wheat flour, while wheat flour is more frequently and to higher degree contaminated with deoxynivalenol.

It is recommended to pursue the problem, because the raw materials of the samples taken for the present study stem from the 2004 and 2005 harvests, which were no fusarium years. That means, cereals of those years were not attacked by fusarium fungi, and the likeliness of formation of mycotoxins was low. The situation may look different in years with strong fusarium attack. Future studies should also look into a possible connection between the degree of grinding of flour and its content in mycotoxins.

6.8

Project 08: Cadmium in cuttlefish products

Office in charge:	LAVES-IFF Cuxhaven
Participating offices:	ILAT Berlin, LGL Oberschleißheim, LHL Wiesbaden, CVUA Freiburg, LVL Rostock, SVUA Arnsberg, LLB Brandenburg

Cuttlefish form part of the marine mollusks and are cephalopods. They have eight (octopods) or ten (decapods) catching arms with suckers. Some 1000 living species are known. Octopods include *Octopus*, while squid (*Loligo sp.* and *Illex sp.*) and *Sepia sp.* are decapods. Cuttlefish is mainly caught in the sea around South East Asia and South Europe.

While fish concentrate only little cadmium in their tissue, cuttlefish, crustaceans and mussels tend to bio-accumulate cadmium. Cadmium levels may be increased in particular in the inner organs of these animals. Cuttlefish products may therefore be contaminated with cadmium, if the raw fish was not thoroughly taken out and cleaned, or if it is sold with the viscera still in.

Cuttlefish rings are characterised by very low cadmium contents. This may be owing to the fact that the cuttlefish is immediately cleaned before it is processed and cut, so that the muscle tissue is not contaminated by the viscera.

The term "cuttlefish product" is used in this project report to generally denominate products derived from the above-mentioned mollusc species.

An assessment of alert notifications through the Rapid Alert System for Food and Feed in the years 2004 and 2005 listed 27 alert notifications about cuttlefish products with high cadmium levels. Most of them referred to consignments which were rejected at border inspection posts. The alert notifications noted cadmium levels of up to 18 mg/kg. The most frequently listed countries of origin were India, Thailand, and Vietnam.

The aim of the project was to get a systematic survey of the situation of contamination in cuttlefish products (except cuttlefish rings) on the German market. The findings were evaluated on the basis of Regulation (EC) No. 466/2001, which fixes a maximum level of cadmium in cuttlefish (without viscera) of 1.0 mg/kg.

A total of 117 cuttlefish products were examined for cadmium. The number of samples per animal species and products and the cadmium concentrations measured are listed in Table P08-1 and graphically represented in Figure P08-1.

It is recognisable that *Sepia* species as a whole display higher cadmium levels than squid (fresh and deep-frozen), and that cuttlefish in sauces and dips as well as octopus show comparatively low levels. This corresponds with the notifications of the rapid alert system. Considering the origin of Sepia samples, which made up the largest portion of the cuttlefish samples with a number of 54, products from Asia (33 samples) were all more contaminated that Sepia products from South Europe (see Figure P08-2).

Some of the laboratories participating in the project also opted for analysing the cuttlefish samples for mercury and lead. These were far below the maximum levels set by Regulation (EC) No. 466/2001, which are 1.0 mg/kg for lead and 0.5 mg/kg for mercury.

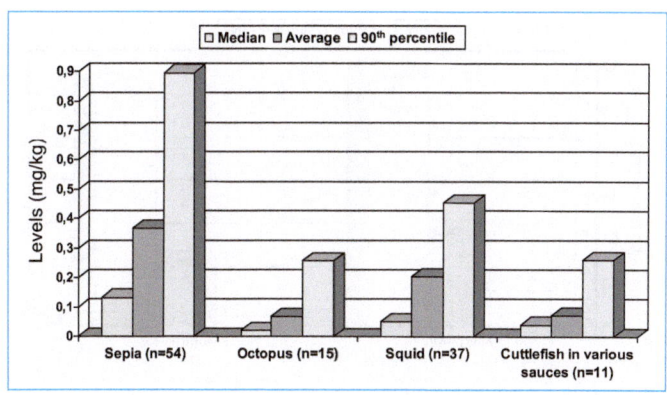

Figure P08-1 Cadmium levels in cuttlefish products (for comparison: the legal maximum level of cadmium in cuttlefish is 1 mg/kg).

Table P08-1 Cadmium levels in cuttlefish products.

Cuttlefish species, -product	Number of samples	Median (mg/kg)	Average (mg/kg)	90th percentile (mg/kg)	Maximum (mg/kg)	Number/proportion (%) > maximum level
Sepia (*Sepia* sp.)	54	0.133	0.368	0.896	3.060	4/7.4
Octopus (*Octopus* sp.)	15	0.020	0.068	0.256	0.324	0
Squid (*Loligo* sp, *Illex* sp.)	37	0.051	0.203	0.455	2.650	2/5.4
Cuttlefish in various sauces/dips	11	0.040	0.071	0.259	0.272	0

Conclusion

The tendency in the findings of this project confirms rapid alert notifications about increased cadmium levels in cuttlefish products mainly from Asia. Yet, a rejection rate of about 5% at the border posts and a maximum finding of cadmium of 3 mg/kg show that border inspection posts are very efficient in rejecting products with extreme cadmium levels. Increased controls and inspection of cuttlefish products will remain an important task of inspectors at border points where imports from third countries enter the EU.

6.9
Project 09: Benzo(a)pyrene in smoked fish

Office in charge: LLB Brandenburg
Participating officers: LAVES-IFF Cuxhaven, CVUA Freiburg, CVUA Stuttgart, LGL Oberschleißheim, CUA Bonn/Köln/Leverkusen, CUA Essen/Wesel/Viersen, LVUA Neumünster, TLLV Jena

Polycyclic aromatic hydrocarbons (PAH) are organic compounds. 250 different substances of that kind are known. They are formed through pyrolysis as products of incomplete burning of materials such as wood, coal, or oil, that means, also during the process of smoking of foodstuffs. Benzo(a)pyrene (BaP) is doubtlessly the most well-known representative of the PAH. That substance is highly carcinogenic and is considered the most important polycyclic aromatic hydrocarbon.

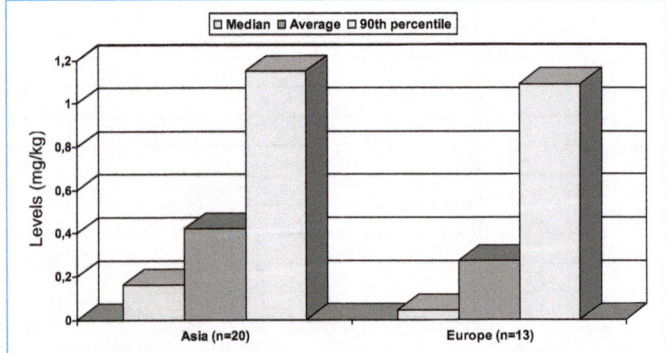

Figure P08-2 Cadmium contamination of cuttlefish products compared by regions of origin. (For comparison: the legal maximum level of cadmium in cuttlefish is 1 mg/kg.).

The generation of PAH depends on the process method. Apart from the way of smoking (direct or indirect), the duration and intensity of the smoke's action on fish are playing a decisive role.

When fish and fish products are smoked directly in small traditional businesses or on market fairs, the crafts people often use traditional ovens, so-called *Altonaer* ovens. It must be assumed that this process leads to a strong contamination of the fish with PAH, because the fish is smoked directly over the wood of beech or alder.

Maximum levels of BaP in smoked fish and fish products were not regulated until the beginning of 2005, neither on the national nor on the EU scale. Only national regulations about aromatic substances fixed a limit of BaP of 1 µg/kg in smoked fish products. That legal limit was a reference for quite a considerable number of complaints raised in the framework of official food control actions about smoked fish products from small businesses.

This situation was the occasion for planning the present monitoring project. The aim was to obtain an overview over the contamination of traditionally smoked fish from small businesses with BaP. Legal maximum levels of BaP in most various foodstuffs had been bindingly set by the EU with Regulation (EC) 208/2005 of 04 February 2005 amending Regulation (EC) 466/2001 with regard to polycyclic aromatic hydrocarbons. With this regulation, the maximum level of BaP in muscle tissue of smoked fish and in smoked fish products (except crustaceans) has been set at 5.0 µg/kg fresh weight (fw) since 01 April 2005.

In total, 176 smoked fish were analysed for BaP, mostly smoked trout or trout fillets as well as mackerel. Some single samples were also drawn from eel, salmon, carp, redfish, halibut, plaice, sheathfish, and whitefish. The maximum level measured was 7 µg/kg and above the legal maximum level. In total, BaP levels were below 1.6 µg/kg in 95% of findings, and below 1 µg/kg in 90% of findings, that is, clearly below the legal maximum level. So, non-compliance with the legal maximum level is an exception.

Conclusion

Project findings show that the vast majority of smoked fish produced in small businesses have BaP levels of less than 1 µg/kg and thus far below the legal maximum level of 5 µg/kg fixed for these products. That maximum level has been repeatedly criticised in literature as too high after it was included in Regulation (EC) No. 466/2001 in 2005, in particular in view of the fact

Table P09-1 Benzo(a)pyrene in smoked fish.

Sample	Number of samples	Median (µg/kg fw)	Average (µg/kg fw)	90th perc. (µg/kg fw)	95th perc. (µg/kg fw)	Maximum (µg/kg fw)	Legal max. level (µg/kg w)
Smoked fish	176	0.1	0.38	1.0	1.6	7.0	5.0

Table P10-1 Residues in various kinds of vegetables.

Vegetable	Number of samples	Samples with residues	Samples with residues above ML	Samples with multiple residues
Leaf vegetables:	134	44 (33%)	8 (6%)	22 (16%)
Basil	4	2 (50%)	1 (25%)	2 (50%)
Savoury	2	1 (50%)	0	1 (50%)
Dill	13	7 (54%)	2 (15%)	3 (23%)
Lamb's lettuce	39	11 (28%)	3 (8%)	5 (13%)
Garden cress	1	0	0	0
Kitchen herbs	4	1 (25%)	1 (25%)	1 (25%)
Parsley leaves	29	9 (31%)	0	3 (10%)
Sage	1	0	0	0
Chives	19	3 (16%)	0	2 (11%)
Spinach	20	10 (50%)	1 (5%)	5 (25%)
Thyme	1	0	0	0
Garden	1	0	0	0
Root vegetables:	75	47 (63%)	3 (4%)	34 (45%)
Carrots	40	15 (38%)	1 (3%)	13 (33%)
Celeriac	35	32 (91%)	2 (6%)	21 (60%)
Total	209	91 (44%)	11 (5%)	56 (27%)

that there had been a lower legal maximum level regulated by the Aromata Regulations for Meat Products before. Given today's technology and know-how of smoking processes, and provided that good processing practices are followed, BaP levels far below the maximum level of 5 µg/kg can be achieved in smoked fish even if the processing is done in small, traditional businesses.

6.10

Project 10: Herbicide residues in certain vegetables

Office in charge:	CVUA Stuttgart
Participating offices:	LUA Bremen, CUA Hagen, LAVES-LI Oldenburg

Although herbicides make up the largest share in the total amount of plant protection products used, there is still only little information about the presence and levels of this class of pesticides in foodstuffs. Herbicides are most widely used in cereals, oil-seed rape, and sugar beet crops, but they are also applied in vegetable crops, in particular such which are machine-harvested. The present project runs over several years and will look into herbicide residues in various vegetable crops with the aim to compile relevant residue data.

In this reporting year, a total of 209 samples of various kinds of vegetables were examined for residues of 61 agreed obligatory herbicide substances, and another maximum 500 active substances of pesticides. Residues of 47 various active substances were found in 44% of the samples, 13 (28%) of the substances being herbicides from the agreed obligatory spectrum of examination.

Residues above legal maximum residue levels were found in 11 vegetable samples (8 leaf vegetables, 3 root vegetables). Three of the cases were attributable to excessive herbicide residues (2x linuron, 1x haloxyfop) in leaf vegetables (2x dill, 1x spinach). Residue findings in the various kinds of vegetables are listed in Table P10-1.

The findings of the project are that active substances of herbicides are quite often found as residues in leaf and root vegetables. Every third plant protection product residue found stemmed from a herbicide (73 out of 229 positive residue findings). Yet, the levels actually found were mostly very low. 71% of the findings were below 0.05 mg/kg food. Any findings of higher levels almost always concerned linuron, which was the herbicide most frequently detected. In two samples of leaf vegetables, linuron was found at even more than 0.5 mg/kg food.

The spectrum of herbicide residues found in leaf vegetables turned out to be much wider than that found in root vegetables. While 11 different herbicides were found in 134 leaf vege-

Table P10-2 Frequency and levels of herbicide residues.

Herbicide	Number of findings	Residues in leaf vegetables	Residues in root vegetables	Levels <0.05 mg/kg	Levels ≥ 0.05 mg/kg
Linuron	53	12	41	35	18
Pendimethalin	4	4	0	3	1
Propyzamide	3	3	0	2	1
Clomazone	2	0	2	2	0
Metobromuron	2	2	0	2	0
Trifluralin	2	2	0	2	0
Clopyralid	1	1	0	1	0
Diuron	1	1	0	1	0
Haloxyfop	1	1	0	0	1
Ioxynil	1	1	0	1	0
Lenacil	1	1	0	1	0
Methabenzthiazuron	1	1	0	1	0
Quizalofop	1	0	1	1	0

table samples, only three herbicides were found in 75 samples of root vegetables (see Table P10-2).

The herbicides detected of the obligatory spectrum of substances are listed in Table P10-2 according to frequency of findings and height of residue level.

Conclusion

The findings of this project show that active substances of herbicides, which account for a large amount of the total application of plant protection products, are also found very frequently as residues in leaf and root vegetables. Every third active substance detected was a herbicide. However, most residues were lower than 0.05 mg/kg. It is conspicuous that the herbicide linuron, which is allowed in Germany only in exceptional cases, was by far the most frequently detected one and had the highest residue levels. Studies in the framework of this project are continued in 2006 to obtain further data on the presence of herbicide residues in various foodstuffs.

7 Survey of foods examined under food monitoring schemes to date

The following table gives a survey over the foods examined in the framework of food monitoring in the years from 1995 to 2005 and the respective years of sampling. Listing of commodity groups and classification of foods with these groups follows the encoding catalogues used in official food control ('ADV catalogues').

Foodstuffs examined in the framework of projects looking into special problems are listed in the table below.

Table 7-1 Market basket foods examined between 1995 and 2005.

Commodity group	Foods examined (year of sampling)
Cheese	Camembert (1999), Emmentaler hard cheese (1995), curd cheese (2000), Gouda (1995), sheep's cheese (1997), goat's cheese (2000)
Butter	Brand butter (1996, 1997)
Eggs	Chicken eggs (2000)
Meat	Duck (2003), goose (2003), rabbit (2003), lamb (2002), beef (2002), ostrich (2002), veal (2001), chicken (2000), turkey (1999), wild boar (1997, 1998)
Viscera	Calf's liver (2001), calf's kidney (2001), lamb's liver (1996), turkey liver (1999), ox liver (1998), ox kidney (2002), pig's liver (1996, 1997), pig's kidney (2001)
Fat tissue	Lamb suet (1996), beef suet (1998), pork flare (1996), fat tissue of wild boar (1997, 1998)
Sausage, meat products	Raw sausage (2005), scalding sausage (2004), quality liver sausage (calves liver) (2000), raw ham/proscuitto ham (2000), black pudding/black pudding slicing sausage (2000), salami sausage (1999, 2005)
Fish, fish products	
Sea fish	Butterfish (2001), shark (2001), halibut (1998), herring (1995, 1996), cod (2002), salmon (2000), rock salmon (1995, 1996), plaice (2001), mock halibut (1998), redfish (2001)
Freshwater fish	trout (1995, 1996, 2005), carp (1997, 1998, 2005)
Fish products	eel smoked (1997), mackerel smoked (1999), tuna tinned (1999)
Crustaceans, molluscs	Crustaceans (1995), edible mussels (1998)
Fats, oils	Olive oil (2000)
Soy products	Tofu (2002)
Cereals	Barley (2001), rice (2000, 2003, 2005), rye (1997, 1998, 2004), wheat (1997, 1998, 1999, 2003)
Cereal products	Puff pastry (2005), bread dough (2005), muesli/cereal bars (2005), wholemeal oat flakes (1999), pasta (2000), wheat bran (2003)
Shelled fruit, oil seed, pulses	Peanuts (1997, 2000, 2004), hazelnuts (2004), linseed (1999, 2005), lentils (2001), almonds (2004), poppy seed (2005), pistachios (1995, 1996, 1998, 1999), sunflower seed kernels (2000), walnuts (2004)
Potatoes, potato products	Potatoes (1998, 2002, 2005), potato puree (2005), potato fritters (2005), croquettes (2005)
Fresh vegetables	
Leaf vegetables	Batavia (1997), celery (1995), Chinese cabbage (2000), oak leaf lettuce (1997), iceberg lettuce (1995, 1996, 1997, 2004), endive (1995, 1996), lamb's lettuce (1995, 1997, 2004), curly kale (1997), head lettuce (1997, 2001, 2004), lollo rosso (1995, 1997), red cabbage (2004), leek (2001, 2004), rucola (2004), spinach (2002, 2005), Savoy cabbage (2000), white cabbage (2003)
Sprouting vegetables	Artichoke (2005), cauliflower (1999, 2003), broccoli (1997, 2005), kohlrabi (1996), asparagus (1998), onions (1999)
Fruiting vegetables	Aubergine (2003), sweet peppers (1999), French beans (1995, 1996, 2002, 2005), cucumbers (1995, 1996, 2000, 2003), melons/cantaloupes (1999), tomatoes (2001, 2004), zucchini (1997)

Commodity group	Foods examined (year of sampling)
Root vegetables	Celeriac (1998), carrots (1998, 2002, 2005), radish (1995, 1996), garden radish (1995, 1996)
Vegetable products	Peas, deep-frozen (2000, 2003), carrot juice (2002), spinach, deep-frozen (1998, 2005), tomato pulp (2000)
Mushrooms, mushroom products	Tinned champignons (2005), shi'itake (2005), cultivated champignons (1999)
Fresh fruit	
Berry fruit	Strawberries (1996, 1998, 2004), currants (1996), table grapes (1995, 1997, 2001)
Pome fruit	Apples (1998, 2001, 2004), pears (1998, 2002, 2005)
Stone fruit	Apricots (1998), nectarines (1998, 2002, 2005), peaches (1998, 2002, 2005), plums (1998), sweet cherries (1998)
Citrus fruit	Clementines (1998), grapefruit (1998), tangerines (2002, 2005), oranges (1996, 1998, 2002, 2005), lemons (1996, 1997, 1998)
Exotic fruit and rhubarb	Pineapple (2004), bananas (1997, 2002), kiwi (1997), papaya (1999), rhubarb (1999)
Fruit products	Apple puree (1995), fruit preparations for dairy products (2001), sour cherries, tinned (2000)
Fruit juices	Pineapple juice (2005), apple juice (1995, 1996, 2005), grapefruit juice (2005), currant nectar (2002), multi-fruit juices (2001), orange juice (1995, 2004), grape must (2005), grape juice, red (2002)
Wine	Quality sparkling wine (2005), red wine (2002), white wine (2001)
Beer	Whole beer (2002)
Honey/sandwich spreads	Honey (2001), nougat paste (1999)
Confectionery, chocolate	Marzipan raw matter (2005), confectionery from other raw matters (2005), chocolate (2002)
Coffee/tea	Green coffee (1999, 2000), roasted coffee (1999), tea unfermented (2002), tea fermented (2002)
Infant food	Ready-meals for infants (2001), milk-free infant food on soy basis (2000), milk powder preparation (1999), fruit puree (2000), infant food on cereal basis (2002), wholemeal cereals-and-fruit preparation (2000)
Spices, herbs	Red pepper (1997), pepper black, white (2002), kitchen herbs (2003)
Drinking water	Mineral water (1999)

Table 7-2 Foodstuffs examined in the framework of special monitoring projects.

Foodstuff	Problem/substance group	Year	Project
Fish, fish products			
Freshwater fish (pike, roach, bream, eel, perch, pike-perch)	Organo-tin compounds	2003	PSM 6
Fish, smoked	Benzo(a)pyrene	2005	9
Herring	Residues and contaminants	2004	9
Mussels/mussel products	Organo-tin compounds and heavy metals	2004	6
Fish tinned with oil (sardine, tuna)	PAH and BTEX	2004	7
Salmon-like fish, cod and perch-like fish, plaice	Mercury in fish from South East Asia	2004	8
Rainbow trout	Polycyclic musk compounds	2004	3
Cuttlefish products	Cadmium	2005	8
Cereals, cereal products			
Bread, crispy products on a cereal basis, pizza, rusk	3-MCPD	2004	10
Breakfast cereals, cereal flakes and cereal products with additives	Deoxynivalenol, zearalenone and ochratoxin A	2004	5
Hard wheat (durum) semolina, pasta, bread	Deoxynivalenol	2003	M 1
Maize flour, maize semolina, cornflakes	Fumonisins	2003	M 3
Rye and wheat flour	Deoxynivalenol, zearalenone and ochratoxin A	2005	7
Fats, oils			
Olive oil, wheat germ oil, maize germ oil	Residues of plant protection products	2003	PSM 3

Foodstuff	Problem/substance group	Year	Project
Potatoes, potato products			
Potatoes	Glycoside alkaloids	2005	3
Vegetables, vegetable products			
Basil, savoury, dill, lamb's lettuce, garden cress, kitchen herbs, parsley, sage, chives, spinach, thyme, garden balm, carrots, celeriac	Herbicides	2005	10
Sweet peppers	Residues of plant protection products	2004	2
Cucumbers	Organo-chlorine compounds, plant protection product residues	2005	6
Tomatoes	Plant protection product residues	2005	5
Fruit, fruit products			
Fruit juices (juices of grapes, apple, pear, oranges, and mixed juices)	Carbendazim	2005	2
Raspberry, currants, gooseberry	Plant protection product residues	2004	1
Raisins, currants, sultanas	Ochratoxin A	2003	M 4
Table grapes red/white	Plant protection product residues	2003	PSM 1
Table grapes red/white	Residues of benzoyl ureas	2003	PSM 2
Food for babies and infants			
Supplementary food for infants on a cereal basis	Deoxynivalenol	2003	M 2
Infant food	Furan	2005	1
Other foodstuffs and combinations of different food groups			
Wheat flour, maize flour, cornflakes, tomato, sweet peppers, carrot, cultivated mushrooms, pear	Residues of chlormequat and mepiquat	2003	PSM 4
Coffee extracts, wine, cocoa powder, spices/spicing products, grape juices, juices for infants	Ochratoxin A	2004	4
Crispbread, butter biscuits, ginger-bread, potato chips prefabricated, potato crisps, coffee roasted	Acrylamide	2004	11
Bouillon and stock products, ready-to-eat meals, sauce powders, soups	Furan	2005	1
Dietary supplements (vitamin, mineral plant extract, and algal preparations)	Heavy metals	2005	4

Glossary

Acrylamide
Acrylamide is formed during production and preparation of foods, both in industrial and private household contexts. Formation of acrylamide is contingent upon the presence of reducing sugars (glucose, fructose) and the amino acid asparagine in the food. These components are present in particular in cereals and potatoes.

Acrylamide has been found carcinogenic and mutagenic in animal experiments. The cancerogenous action is believed to root in a genotoxic mechanism. Yet, current data are not sufficient to allow a final assessment of the risk potential of acrylamide for humans.

Acaricides
Agents applied to kill mites.

Aflatoxins
Aflatoxins are metabolic products of mould fungi. Formation of aflatoxins is promoted by warmth and humidity. Aflatoxins include the chemically related compounds aflatoxin B1, B2, G1, G2, and M1. They are acutely toxic and have been found to cause carcinomas of the liver in various animal species through a genotoxic mechanism. Formation of liver carcinomas in humans is also brought into connection with the hepatitis-B virus. To prevent risks to human health by foodstuffs contaminated with aflatoxins, the EU has established maximum levels, namely 2 µg/kg for aflatoxin B1, 4 µg/kg for the sum of all aflatoxins, and 0.05 µg/kg for M1 in milk.

Average value
The average is a statistical measure serving to characterise data. The present report always uses the arithmetical average, calculated as the sum of all measuring values divided by their number.

Benzo(a)pyrene
Benzo(a)pyrene is the most well-known representative of polycyclic aromatic hydrocarbons (PAH; see also definition of PAH) and is considered the major substance of that group. The substance is highly carcinogenic and mutagenic.

Contaminants
Each substance which is not intentionally added to a food, but is present in a food as a result of its production (including treatment methods in cropping, animal husbandry, and veterinary medicine), conversion, preparation, processing, packing, transport, and storage, or as a result of environmental effects. The term does not include foreign bodies such as residues of insects, hair of rodents, or others.

Contamination
Contamination of foods with undesirable substances.

Degree of contamination
To describe the degree of contamination of a product, one proceeds from the proportion of samples found with contaminant levels above the established maximum levels (ML). This share is evaluated according to the following scale:

Evaluation	Share > ML (in %)
1 – no contamination	share = 0
2 – low	0 < share <= 5
3 – medium	5 < share <= 10
4 – enhanced	10 < share <= 15
5 – high	share > 15

Criteria are similar for evaluating the actual height of levels found, or the share of samples with residues or contaminants above detection levels.

Deoxynivalenol
Deoxynivalenol (DON) is a metabolic product of mould fungi. It counts among the fusarium toxins. DON may occur in all cereal species, but occurs primarily in maize and wheat. While it is neither mutagenic nor carcinogenic, it has often an acute toxic effect, causing nausea, diarrhoea, and skin irritation upon consumption of contaminated food. It may also disturb the immune system, making humans more susceptible to infectious diseases.

Elements
In the framework of this food monitoring, the term "elements" designates the heavy metals and semi-metals, such as arsenic and selenium (see also definition of "Heavy metals").

Frequently quantified substances
The meaning of the criterion 'frequent' depends on the substance group in question. Here, the term frequent was applied when residues of plant protection products or mycotoxins were quantified in more than 10% of samples. With organic contaminants and elements, 'frequently quantified' meant in more than 50% of all samples.

Fungicides

Substances preventing or impairing the growth of micro-fungi, for instance mould fungi.

Furan

Furan is a very small molecule, a cyclic five-membered ring without an attached chain, with oxygen as a hetero atom in the ring. It is very volatile, with a boiling point of 31 °C, and has an ether-like smell. Furan may be formed in foodstuffs in the course of the Maillard reaction when carbohydrates are heated. It is also formed when ascorbic acid, amino acids, or multi-unsaturated fatty acids are heated. Furan levels are climbing in particular when food products are roasted (coffee beans), or heated in hermetic systems, such as infant food or ready-to-eat meals. After furan was ranked as carcinogenic in animal tests after an NTP study in the USA in 1993 [NTP, 1993: Toxicology and carcinogenesis studies of furan (CAS No. 110-00-9) in F344/N rats and B6C3F1 mice (gavage studies), NTP Technical Report No. 402., U.S. Department of Health and Human Services, Public Health Service, National Institutes of Health, Research Triangle Park, NC, 1993], the WHO also ranked furan as potentially carcinogenic in humans in 1995. Its exact action in the human body is still not clarified, though. It is assumed that furan, similar to benzene, is rendered ineffective through the cytochrome-P450 system. The intermediary product cis-2-butene-1,4-dial can interact with the DNA.

Glycoside alkaloids

The poisonous glycoside alkaloids solanine and chaconine naturally occur in small amounts in potatoes. They may be formed at increased levels in greened, germinating, or damaged potatoes, consumptions of which may lead to disturbances of the nervous and gastro-intestinal systems. Solanine and chaconine are mainly present in the peel of potatoes. They are not destroyed by cooking, but largely migrate to the boiling water, which should be poured away. A total alkaloid content (that is, the sum of solanine and chaconine) in potatoes of up to 200 mg/kg is considered harmless. The Joint FAO/WHO Expert Committee on Food Additives has evaluated a glycoside alkaloid level of 20 to 100 mg/kg in potatoes as normal[1].

Heavy metals

Heavy metals are such with a density of 4.5 g/cm³ and more. Lead, cadmium, mercury, and tin are well-known representatives. Nickel, thallium and zinc sometimes also play a role in the contamination of foodstuffs. Heavy metals can enter foods through the air, water, and soil, but also in the course of treatment and processing. Evaluation of heavy metal levels in foodstuffs is based on the provisions of the Contaminants Regulation (EC) No. 466/2001, on the German Regulations on Harmful Substances as regards mercury, and also on the German Regulations on Maximum Residue Levels as regards copper and mercury.

Herbicides

Weed control products

Histamine

Histamine is a biogenic amine which is formed through bacterial decomposition of the amino acid histidine. Biogenic amines are inherent to fish, meat, cheese, wine, and various kinds of vegetables. Levels are generally very low in unprocessed foodstuffs, but may sharply increase through microbiological processes, such as fermentation and maturing, or during storage. Therefore, fermented foodstuffs, such as *Sauerkraut*, beer, wine, and cheese, in particular long-cured kinds, and sausage – namely salami and raw ham – may contain high levels of histamine. The same holds for improperly stored perishable foodstuffs, such as tuna and mackerel.

Humans take in an average 4 milligram histamine per day with the diet. Normally, the human body is also able to metabolise higher levels of histamine. In case of an histamine intolerance, however, 15 to 30 microgram histamine may be enough to cause symptoms. This amount would be contained, for instance, in a glass of red wine, or a small piece of old Gouda. In higher concentrations (>1000 mg per meal), histamine has adverse effects on health, causing skin irritation, headaches, nausea and diarrhoea, up to life-threatening conditions such as dyspnoea.

More information including on histamine contents in some foodstuffs may be obtained under http://www.was-wir-essen.de/infosfuer/histamin_intoleranz_6494.php.

HMF (5-hydroxymethyl furfural)

This is a breakdown product of sugar and carbohydrates which may be formed under the influence of heat and through improper storage. It is suspected to have mutagenic and carcinogenic effect. HMF was detected in a wide variety of heat-treated products, such as milk, fruit juices, dried fruit, coffee, spirits, and honey. Higher levels of HMF are mainly found in foodstuffs containing more saccharose and fructose. Dried fruit, for instance, was found to contain concentrations between 10 and 100 mg/kg, while dried plums and plum jam even carried up to 2 g HMF per kg product[2]. Obviously, fruit acids are also of particular importance for HMF formation, apart from predominant sugars and amino acids. Commercial sorts of coffee were found to contain HMF concentrations between 300 and 2000 mg/kg, and more than half of the concentrations at that time were found to be higher than 1 g/kg.

Insecticides

Insect control products

Limit of detection

Undesirable substances in foods, such as pesticide residues, are analysed using complicated and expensive methods and equipment. Yet, there is a bottom limit for the qualitative detection of a substance. If the amount of substance present in a food is less than that limit, the substance can no longer be detected. The

---

[1] Summary of Evaluations Performed by the Joint FAO/WHO Expert Committee on Food Additives, http://www.inchem.org/documents/jecfa/jeceval/jec_2180.htm, http://www.inchem.org/documents/jecfa/jeceval/jec_399.htm, 1992

[2] Murkovich M. und Pichler N. (2006) Analysis of 5-hydroxymethylfurfural in coffee, dried fruits and urine. Mol. Nutr. Food Res. 50:842-846.

minimum amount of a substance that must be present for that substance to be detected is termed the "limit of detection" (see also the definition of "limit of quantification").

Limit of quantification

The smallest quantity of a substance that can safely be determined ("quantified") in another substance is called the limit of quantification. That limit depends on the method and techniques used and is higher than the respective limit of detection. The present report usually does not make a difference between the two limits, and all residues which are below the limit of quantification are classified as "not detectable". This is a minor inaccuracy which is accepted to the end of enhancing easy readibility of this report (*cf.* the term "limit of detection").

Maximum level

Maximum levels designate limits of plant protection products and contaminants in or on foods which are fixed as maximum permissible in EU legislation and must not be exceeded when foods are commercially marketed. They are fixed as low as possible on the basis of internationally recognised, strictly scientific standards, and never lie above what is toxicologically tolerable.

Responsibility for compliance with these maximum levels lies, in the first place, with the EU-resident producer or, in case of imports from third countries, the EU-resident importer of foods. Official food control bodies of the Member States make random tests of foods for compliance with EU maximum levels.

The term Maximum Residue Level (MRL) is used in Germany in legal regulations about residues of plant protection products in and on plants, such as in the Regulations on Maximum Residue Levels (*Rückstands-Höchstmengenverordnung – RHmV*). Its use and meaning correspond with the term maximum level used in EU legislation.

Median

The median is the particular value which cuts in two halves a series of measuring values arranged by size. That means half of the measuring values lie below the median, the other half above the median.

The median is preferably used to characterise asymmetric distributions, such as of substance concentrations in foods. When all samples of a measuring series are considered, that is, including such without quantifiable residues, a median can only be stated if at least half of all samples carried measurable residues. Otherwise, the median would be zero by definition.

Metabolites

Metabolites are conversion products of chemical compounds produced as a result of chemical or metabolic processes.

Musk compounds

Cheap and easily produced nitro musk compounds, such as musk xylene and musk ketone, were first used as synthetic musk compounds to substitute natural musk. As the toxic risks related with these substances became known, their use was strongly restricted. As a result, concentration of these substances in the environment as well as in food samples has noticeably declined over the past few years.

To substitute nitro musk compounds, polycyclic musk compounds were elected, believing that these are safe in ecological and toxicological respect. Yet, it has been meanwhile found out that some substances of that substance group, with galaxolide and tonalide in the first line, can also accumulate in the aquatic food chain, so that residues are found both in sea fish and in freshwater fish. Given that some polycyclic musk compounds may also have toxic effects, these substances should remain part of monitoring programmes and control measures in the time ahead. There are not yet any legal regulations on how to evaluate these substances.

Mycotoxins

Mycotoxins are toxic substances formed as products of metabolic activity in some mould fungi. They have very different chemical structures and may be formed on foodstuffs and feed. They are produced either by plant-pathogenic or by apathogenic fungi during the growth of crop plants, or by storage fungi during storage and processing of commodities. Warmth and humidity promote formation of mycotoxins. Toxicologists say that mycotoxins are among the most toxic substances to be found in food and feed. Well-known mycotoxins are the aflatoxins, fusarium toxins, and ochratoxin A (OTA).

Nitrate, nitrite, nitrosamines

Nitrate is a substance naturally occurring in soil. Plants need it for their growth, which is why it is added to the soil as fertiliser. Excessive fertilisation can lead to very high nitrate levels in plants. But the nitrate level also depends on other influences, such as plant species, the time of harvest, weather, and climatic conditions. Light is playing a decisive role. Nitrate levels are usually higher in the months with less light.

Nitrate may be reduced to nitrite in the human gastro-intestinal tract, and nitrite might react with certain protein substances to form nitrosamines. Nitrosamines, again, have been found carcinogenic in animal studies.

Evaluation of nitrate levels is based on the following classification:

Classification of levels	Criterion underlying the classification
1 – very low	Average level <= 0.10 * RV
2 – low	0.10 * RV < average level <= 0.25 * RV
3 – medium	0.25 * RV < average level <= 0.50 * RV
4 – increased	0.50 * RV < average level <= 0.75 * RV
5 – high	Average level > 0.75 * RV

In this classification, the level is expressed as the average of levels measured and referred to a reference value (RV). Here, the reference value is the legal maximum level. If there is no legal maximum level, a reference value is defined depending on the potential contamination of the fruit or vegetable in question in the following way:

Potential contamination	Fruit/vegetables	Reference value (mg/kg)
Minor nitrate contamination	Cauliflower, peas, cucumber, sweet peppers, tomato, French beans, potatoes, onions, different kinds of fruit	500
Medium nitrate contamination	Carrots, celeriac, cabbage, leek, rhubarb	1000
High nitrate contamination	Leaf and head lettuce, Chinese cabbage, spinach, kohlrabi, radish, red turnip beet, celery	4000

Ochratoxin A

Metabolic product of mould fungi which damages the liver and kidneys. Production of ochratoxin A is promoted by warmth and humidity. Ochratoxin A mostly occurs in cereals, coffee beans, and oil seed. It may be found in food of animal origina, for instance in milk, when the animals were fed with ochratoxin-A-containing feed.

Organo-chlorine compounds (persistent chlorinated hydrocarbons)

Organo-chlorine compounds are persistent, hardly degradable substances. Because of their persistence they may be present as residues in foodstuffs. Chlorinated hydrocarbons are, for instance, HCB, DDT, and PCB. Apart from DDT isomers, their metabolites DDD and DDE are also often found.

Patulin

Patulin is a metabolic product of mould fungi in fruit. It is found in fruit products in particular when the raw material fruit was of deficient quality. In animal tests, patulin caused weight loss and damage to the gastro-intestinal mucosa when administered in larger amounts and over a longer period of time. There were also signs of possible genotoxic effect.

PCB (Polychlorinated biphenyls)

PCB used to be frequently applied for industrial purposes (e. g. technical oils, heat transmitters, plasticisers). PCB is a mixture of many single compounds with different degrees of chlorination (congeners). PCBs are poorly degraded and enter the human food chain via the soil, water and animal feeds. PCB 138, PCB 153, and PCB 180 are compounds frequently found in food of animal origin.

Percentile

Similar to the median, percentiles are values dividing into two series of measurements arranged according to size. For instance, the 90th percentile represents the value below which 90% of the measurements are found, while 10% of measurements lie above this value.

Plant protection products (PPP)

Plant protection products are used in agricultural production to protect plants from harmful organisms and diseases, thus safeguarding crop commodities from spoilage and securing yields. Consumers are effectively protected by legal regulations about the authorisation of plant protection products and residue controls. Authorisation procedures ensure that plant protection products do not have harmful effects on the health of man and animal if they are properly used. Excessive residue levels are primarily a result of improper use of such products.

Plant protection products are divided by function into insecticides, fungicides, herbicides, acaricides, and others.

Polycyclic aromatic hydrocarbons (PAH)

PAH is a collective term for several hundred individual components of condensed, aromatic hydrocarbons. PAH are formed as unwanted by-products through incomplete combustion processes and heat treatment under airtight conditions. That means, they may also be formed in foodstuffs when these are heat-treated, dried or smoked and come into direct contact with combustion residues. Some of the PAH components are carcinogenic, or harmful to the human organism in different ways. Most have a very penetrant smell. The most well-known PAH with adverse effects on health is benzo(a)pyrene (BaP). This compound is often cited as a standard substance for the analysis and toxicological evaluation of PAH contamination.

An extended toxicological evaluation may also include the so-called "heavy" PAHs, which include, apart from benzo(a)pyrene, the compounds dibenzo(a,h)anthracene, benzo(b)fluranthene, benzo(k)fluranthene, benzo(ghi)perylene, and indeno(1,2,3,c,d)pyrene.

Quantified levels

If the concentration of a substance is high enough to be reliably determined by the analytical method chosen, this concentration found (measured value) is referred to as a quantified level (*cf.* "Limit of quantification").

Rapid Alert System for Food and Feed (RASFF)

Immediate action is needed if foodstuffs or feed are contaminated or harbour any other risk for consumers. The European Union has therefore established a Rapid Alert System for Food and Feed (RASFF) for rapid relay of information within Community. The legal basis of the RASFF is Article 50 of Regulation (EC) No. 178/2002. The national contact point of the RASFF in Germany is the Federal Office of Consumer Protection and Food Safety *(Bundesamt für Verbraucherschutz und Lebensmittelsicherheit, BVL)*. From the national side, BVL receives alert notifications by the *Bundesländer* (states) about products which pose a risk to consumers. The alert notifications are checked and completed according to a prescribed procedure and then passed on to the Member States of the European Union. From the Community side, the BVL has the task to inform the authorities of the *Bundesländer* about the alert notification received via RASFF from other Member States of the EU.

Recommendation of the co-ordinated Community monitoring programme

The co-ordinated Community monitoring programme is based on recommendations by the European Union to its Member States concerning compliance with maximum residue levels of plant protection and pest control products in and on cereals and certain other products of plant origin. Compliance with these recommendations makes test results representative and comparable. Recommendations for 2005 have been published in the Official Journal of the European Communities No. L 61/31 of March 8, 2005, under the title "Commission Recommendation of 01 March 2005 concerning a co-ordinated Community monitoring programme for 2005 to ensure compliance with maximum levels of pesticide residues in and on cereals and certain other products of plant origin, and concerning national monitoring programme of the year 2006".

Statement of residue levels

Residue levels are stated as mg/kg (milligram per kilogram) or as µg/kg (microgram per kilogram). In beverages, the unit mg/l is used.

1 mg/kg (or mg/l) means that there is a residue of a thousandth of a gram (a milligram) of a certain substance in one kilogram (or one litre) of a food product. 1 µg/kg means that there is a residue of one millionth of a gram of a substance in one kilogram of a food.

Ubiquitous

Present everywhere

Zearalenone

Zearalenone is a metabolic product of fusarium fungi (*Fusarium graminearum*). It has estrogenous and anabolic effects. Its acute toxicity is low. Zearalenone is mainly formed in maize and other cereals with humid conditions and low temperatures. A maximum level of 50 µg/kg was established for cereals in 2004.

Addresses of the ministries responsible for food monitoring activities, Federal Government office in charge

Federal ministry

Bundesministerium für Ernährung, Landwirtschaft und Verbraucherschutz
Postfach 14 02 70
53107 Bonn
Telefax: 01888/529 4262
E-Mail: 313@bmelv.bund.de

Federal office in charge

Bundesamt für Verbraucherschutz und Lebensmittelsicherheit, Dienstsitz Berlin,
Postfach 10 02 14
10562 Berlin
Telefax: 030/18444 89999
E-Mail: poststelle@bvl.bund.de

State ministries

Ministerium für Ernährung und Ländlichen Raum Baden-Württemberg
Kernerplatz 10
70182 Stuttgart
Telefax: 0711/126 2255
E-Mail: poststelle@mlr.bwl.de

Bayerisches Staatsministerium für Umwelt, Gesundheit und Verbraucherschutz
Rosenkavalierplatz 2
81925 München
Telefax: 089/9214 3505
E-Mail: poststelle@stmugv.bayern.de

Senatsverwaltung für Gesundheit, Umwelt und Verbraucherschutz
Oranienstr. 106
10969 Berlin
Telefax: 030/9028 2060
E-Mail: poststelle@sengsv.verwalt-berlin.de

Ministerium für Ländliche Entwicklung, Umwelt und Verbraucherschutz des Landes Brandenburg
Postfach 60 11 50
14411 Potsdam
Telefax: 0331/866 4069
E-Mail: poststelle@mlur.brandenburg.de

Senator für Arbeit, Frauen, Gesundheit, Jugend und Soziales
Bahnhofplatz 29
28195 Bremen
Telefax: 0421/361 4808
E-Mail: veterinaerwesen@gesundheit.bremen.de

Behörde für Soziales, Familie, Gesundheit und Verbraucherschutz
Amt für Gesundheit und Verbraucherschutz
Billstr. 80a
20359 Hamburg
Telefax: 040/428 37 2401
E-Mail: inga.ollroge@bsg.hamburg.de

Hessisches Ministerium für Umwelt, ländlichen Raum und Verbraucherschutz
Mainzer Str. 80
65189 Wiesbaden
Telefax: 0611/4478 9771
E-Mail: poststelle@hmulv.hessen.de

Ministerium für Ernährung, Landwirtschaft, Forsten und Fischerei Mecklenburg-Vorpommern
Paulshöher Weg 1
19061 Schwerin
Telefax: 0385/588 6025
E-Mail: lm-presse@mvnet.de

Niedersächsisches Ministerium für den ländlichen Raum, Ernährung, Landwirtschaft und Verbraucherschutz
Calenberger Str. 2
30169 Hannover
Telefax: 0511/120 2385
E-Mail: poststelle@ml.niedersachsen.de

Ministerium für Umwelt, Naturschutz, Landwirtschaft und Verbraucherschutz des Landes Nordrhein-Westfalen
Schwannstr. 3
40476 Düsseldorf
Telefax: 0211/456 6388
E-Mail: poststelle@munlv.nrw.de

Ministerium für Umwelt, Forsten und Verbraucherschutz Rheinland-Pfalz
Kaiser-Friedrich-Str. 1
55116 Mainz
Telefax: 06131/164 608
E-Mail: poststelle@mufv.rlp.de

Ministerium für Justiz, Gesundheit und Soziales
Postfach 10 24 53
66024 Saarbrücken
Telefax: 0681/501 3335
E-Mail: poststelle@soziales.saarland.de

Sächsisches Staatsministerium für Soziales
Albertstr. 10
01097 Dresden
Telefax: 0351/564 5770
E-Mail: poststelle@sms.sachsen.de

Ministerium für Gesundheit und Soziales des Landes Sachsen-Anhalt
Turmschanzenstr. 25
39114 Magdeburg
Telefax: 0391/567 4688
E-Mail: poststelle@ms.lsa-net.de

Ministerium für Landwirtschaft, Umwelt und ländliche Räume des Landes Schleswig-Holstein
Mercatorstraße 3
24106 Kiel
Telefax: 0431/988 5246
E-Mail: poststelle@MLUR.landsh.de

Thüringer Ministerium für Soziales, Familie und Gesundheit
Postfach 90 03 54
99106 Erfurt
Telefax: 0361/379 8850
E-Mail: poststelle@tmsfg.thueringen.de

List of the state (*Länder*) laboratories responsible for food monitoring

Baden-Württemberg

Chemisches und Veterinäruntersuchungsamt, Freiburg

Chemisches und Veterinäruntersuchungsamt, Karlsruhe

Chemisches und Veterinäruntersuchungsamt, Sigmaringen

Chemisches und Veterinäruntersuchungsamt, Stuttgart, Sitz Fellbach

Bavaria

Bayerisches Landesamt für Gesundheit und Lebensmittel-sicherheit, Erlangen

Bayerisches Landesamt für Gesundheit und Lebensmittel-sicherheit, Dienststelle Oberschleißheim

Berlin

Berliner Betrieb für Zentrale Gesundheitliche Aufgaben (BBGes) – Institut für Lebensmittel, Arzneimittel und Tierseuchen (ILAT)

Brandenburg

Landesamt für Verbraucherschutz und Landwirtschaft, Laborbereich Potsdam

Landesamt für Verbraucherschutz und Landwirtschaft, Laborbereich Frankfurt/Oder

Bremen

Landesuntersuchungsamt für Chemie, Hygiene und Veterinärmedizin

Hamburg

Institut für Hygiene und Umwelt
Hamburger Landesinstitut für Lebensmittelsicherheit, Gesundheitsschutz und Umweltuntersuchungen

Hesse

Landesbetrieb Hessisches Landeslabor, Standort Kassel

Landesbetrieb Hessisches Landeslabor, Standort Wiesbaden

Mecklenburg-Western Pomerania

Landesveterinär- und Lebensmitteluntersuchungsamt Mecklenburg-Vorpommern, Rostock

Lower Saxony

Niedersächsisches Landesamt für Verbraucherschutz und Lebensmittelsicherheit, Lebensmittelinstitut Braunschweig

Niedersächsisches Landesamt für Verbraucherschutz und Lebensmittelsicherheit, Lebensmittelinstitut Oldenburg

Niedersächsisches Landesamt für Verbraucherschutz und Lebensmittelsicherheit, Institut für Fischkunde Cuxhaven

Niedersächsisches Landesamt für Verbraucherschutz und Lebensmittelsicherheit, Veterinärinstitut Hannover

Niedersächsisches Landesamt für Verbraucherschutz und Lebensmittelsicherheit, Veterinärinstitut Oldenburg, Außenstelle Stade

North Rhine-Westphalia

Chemisches und Lebensmitteluntersuchungsamt der Stadt Aachen

Staatliches Veterinäruntersuchungsamt Arnsberg

Chemisches Untersuchungsamt der Stadt Bochum

Amt für Umweltschutz und Lokale Agenda der Stadt Bonn

Chemisches und Veterinäruntersuchungsamt Ostwestfalen-Lippe, Detmold

Chemisches und Lebensmitteluntersuchungsamt der Stadt Dortmund

Chemisches Lebensmitteluntersuchungsamt der Stadt Düsseldorf

Chemisches und Geowissenschaftliches Institut der Städte Essen und Oberhausen

Chemisches Untersuchungsamt der Stadt Hagen

Chemisches Untersuchungsamt der Stadt Hamm

Institut für Lebensmittel- und Umweltuntersuchungen der Stadt Köln

Chemisches Untersuchungsinstitut der Stadt Leverkusen

Amt für Verbraucherschutz des Kreises Mettmann

Institut für Lebensmitteluntersuchungen und Umwelthygiene für die Kreise Wesel und Kleve, Moers

Chemisches Landes- und Staatliches Veterinäruntersuchungsamt, Münster

Gemeinsames Chemisches und Lebensmitteluntersuchungsamt für den Kreis Recklinghausen und die Stadt Gelsenkirchen in der Emscher-Lippe-Region (CEL), Recklinghausen

Chemisches Untersuchungsinstitut Bergisches Land Wuppertal

Rhineland-Palatinate
Landesuntersuchungsamt Rheinland-Pfalz
Institut für Lebensmittel tierischer Herkunft Koblenz

Landesuntersuchungsamt Rheinland-Pfalz
Institut für Lebensmittelchemie und Arzneimittelprüfung Mainz

Landesuntersuchungsamt Rheinland-Pfalz, Institut für Lebensmittelchemie Speyer

Landesuntersuchungsamt Rheinland-Pfalz, Institut für Lebensmittelchemie Trier

Saarland
Landesamt für Soziales, Gesundheits- und Verbraucherschutz Saarbrücken

Saxony
Landesuntersuchungsanstalt für das Gesundheits- und Veterinärwesen Sachsen, Standort Chemnitz

Landesuntersuchungsanstalt für das Gesundheits- und Veterinärwesen Sachsen, Standort Dresden

Landesuntersuchungsanstalt für das Gesundheits- und Veterinärwesen Sachsen, Standort Leipzig

Sachsen-Anhalt
Landesamt für Verbraucherschutz Sachsen-Anhalt, Standorte Halle und Stendal

Schleswig-Holstein
Landeslabor Schleswig-Holstein, Neumünster
Landeslabor Schleswig-Holstein, Außenstelle Kiel I

Thuringia
Thüringer Landesamt für Lebensmittelsicherheit und Verbraucherschutz, Standort Bad Langensalza

Thüringer Landesamt für Lebensmittelsicherheit und Verbraucherschutz, Standort Erfurt

Thüringer Landesamt für Lebensmittelsicherheit und Verbraucherschutz, Standort Jena